Energy Dissipation

A. Tramer   Ch. Jungen   F. Lahmani

Energy Dissipation in Molecular Systems

With 85 Figures and 16 Tables

 Springer

André Tramer

pl. de la Fontaine (Mulleron)
91640 Janvry, France
e-mail: andre.tramer@wanadoo.fr

Christian Jungen

Université de Paris Sud
Laboratoire Aimé Cotton CNRS
91405 Orsay, France
e-mail: christian.jungen@lac.u-psud.fr

Françoise Lahmani

Université de Paris Sud
Laboratoire de Photophysique Moléculaire CNRS
91405 Orsay, France
e-mail: francoise.lahmani@ppm.u-psud.fr

ISBN 978-3-642-06409-8   e-ISBN 978-3-540-26566-5

Springer is a part of Springer Science+Business Media
springeronline.com

Cover design: *design & production* GmbH, Heidelberg

Printed on acid-free paper   3141/JVG   5 4 3 2 1 0

Preface

The scope of this book is the analysis of experimental data on the energy relaxation in molecular systems using as a basis the standard theory of radiationless transitions. The main principles of this theory were established in the 1960s and early 1970s, when only a small amount of data obtained by traditional spectroscopic methods was available. Since this time, a rapid development of techniques (short-pulse and narrow-band lasers, supersonic nozzles and molecular beams, multiple-resonance techniques, the single-molecule spectroscopy) allowed the experimental studies of a large variety of molecular systems with practically unlimited time and energy resolution. In the same time, the progress of the theory opened new fields for calculations and simulation of molecular processes.

The major part of important chemical and biological processes takes place in condensed phases, where essentially intramolecular processes cannot be separated from the intermolecular interactions involving the energy exchange between the molecular system and its heat bath.

In order to separate the effects due to intrinsic properties of the molecule and those induced by the external perturbation, the studies of molecules "isolated" (i.e., collision-free during a sufficiently long time) in the gas phase and in the supersonic expansions have been widely developed. On the other hand, a better understanding of intermolecular interactions was attained in the studies of simple model systems such as two-body collisions in the gas phase, 1:1 molecular complexes, molecules trapped in rare-gas clusters and crystals. These studies allowed us to see how different intramolecular processes are modified by the environment effects.

We have at our disposal an enormous amount of data concerning the vibrational energy redistribution and relaxation, electronic energy relaxation, electron and proton transfer in a wide variety of molecular systems involving molecules isolated or dissolved in fluid or rigid solvents.

It seems to us that it is a good moment for a general discussion of key problems of the molecular dynamics based on the analysis of these data. We try to prepare this discussion by description, in the same language, of different photophysical processes.

We do not discuss the recent developments of the theory; our approach may be considered as complementary with respect to review papers and books (cf. Sect. 1.1) focused on the theoretical aspects of photophysics and photochemistry. In any way, the basic concepts of the theory remain unchanged and only oversimplified schemes of level-coupling mechanisms used in its childhood must be replaced by a more detailed description of the coupling pattern accounting for a hierarchy of coupling chains corresponding to different time scales of sequential decay processes.

The outlines of the theory of nonradiative transitions are presented in the simplest form in Chap. 1. In Chap. 2, we review briefly principal experimental techniques with emphasis on the time- and energy-resolution that may be attained in experiments involving the pumping and probing populations of molecular excited states.

In the next chapters, we discuss the processes of the intramolecular redistribution of the vibrational energy in isolated molecules (Chap. 3), vibrational and rotational relaxation induced by the environment (Chap. 4) and electronic relaxation of isolated and medium-perturbed molecular systems (Chap. 5). At last, we briefly treat, in Chap. 6, specific processes of relaxation involving intra- and intermolecular electron or proton (hydrogen atom) transfer as examples of the simplest monomolecular photochemical processes. The review of experimental data is limited to a small number of small- and intermediate-size molecules that may be considered as representative of a larger class of molecular systems. The supermolecular ensembles (crystals and polymers) as well as biological systems are not treated here.

This book is not a manual introducing systematically the fundamental notions of a domain. It does not pretend either to be a monograph giving an extensive review of data and its complete bibliography. It may be rather considered as a tourist guide of the country of photophysics, focusing attention on several of the most famous monuments. The choice of these masterpieces is, to some extent, subjective.

We focus our attention on the physical meaning rather than on a rigorous development of the quantum-mechanical formalism. We tried to get this book accessible to everybody having a knowledge of basic notions of the quantum mechanics and molecular spectroscopy as somebody starting graduate studies of molecular physics or physical chemistry. We hope that it may be useful for some people actively working in this field.

The authors are highly indebted to Alberto Beswick, Claudine Crépin-Gilbert, Irena Deperasinska, Christophe Jouvet, Philippe Millié, François Piuzzi and Jerzy Prochorow for discussions, suggestions and critical remarks. Special thanks are due to René Voltz and Anne Zehnacker-Rentien, whose roles were particularly important.

Orsay, January 2005   André Tramer
                                                       Christian Jungen
                                                       Françoise Lahmani

Contents

1

The Basic Notions

1.1 Introductory Remarks

The problem of nonradiative relaxation of electronically excited molecules is as old as the experimental studies in the field, called still 60 years ago luminescence, and separated from the gas-phase spectroscopy of atoms and small molecules (cf. [1]). These studies—limited in practice to condensed phases—showed that the major part of chemical compounds are nonluminescent and that the important fraction of radiation energy absorbed by the luminescent ones is dissipated in the medium without emission of light. Such a deactivation process seemed not to occur for isolated gas-phase atoms and diatomics in the absence of collisional quenching. It was, therefore, not evident whether the nonradiative relaxation is the intrinsic property of large molecules or is induced by the environment effects.

In the early 1960s, it was shown that radiationless processes occur in isolated polyatomic molecules in the gas phase. It was shown, for instance, that the fluorescence yield of the collision-free benzene molecule in the gas phase is nearly the same as in its solutions in aliphatic solvents [2] and that nonradiative processes occur in an isolated molecule as small as glyoxal $C_2H_2O_2$ [3].

The theory of nonradiative transitions was developed at the same time; after several pioneering works [4–7], the publication in 1968 of the Bixon and Jortner paper [8] may be considered as its birthdate. The further progress of the theory is described in a number of review papers published in the 1970s (cf. [9–13] and in recent books [14, 15].

In this chapter we give—in a simplified form—some of the basic formulae necessary for the analysis of experimental data concerning the time evolution of excited electronic and vibrational states of molecules, isolated or interacting with their environment, the attention being focused on the former ones. The outlines of the traditional theory are presented with a few minor modifications. The semiclassical approach (the quantum theory for molecules and a classical description of radiation fields) is adopted.

The time evolution of an isolated molecule is determined by the fine details of its level patterns. The information about its dynamics may be thus deduced either from the *time-resolved* monitoring of molecular systems prepared by a short-pulse excitation to a well-defined level or set of levels or from *energy-resolved* spectroscopic studies of its fine structure. We will discuss separately, after some introductory remarks (Sect. 1.2), the information from the absorption spectroscopy (Sect. 1.3) and from the time-resolved experiments (Sect. 1.4), summarized in Sect. 1.5. At last, in Sect. 1.6, we introduce some general notions concerning the dependence of transition probabilities on the energy or momentum gap between initial and final state of the system.

1.2 Excited Molecular States

1.2.1 Stationary and Quasi-Stationary States

Stationary states are normally envisaged for isolated molecules in the absence of interactions with material or radiation-field environments. The state vectors are then of the form:

$$\varphi(t) = |\varphi\rangle \exp[-iE_\varphi t/\hbar]. \tag{1.1}$$

The lowest E_φ value is usually taken as the origin of the energy axis, $E_0 = 0$, and attributed to the ground state $|0\rangle$.

When the molecule interacts with the radiation field, the ground state remains stationary but the excited states become unstable due to the spontaneous light emission: transitions to lower states $|\varphi'\rangle$. The energy difference $E_\varphi - E_{\varphi'}$ is transformed in a photon $\hbar\omega_{\varphi\varphi'}$ (Fig. 1.1). These processes are properly described in an exact quantal treatment considering the molecular and radiation-field Hamiltonians together with a light-matter interaction term. In the simpler semiclassical treatment, it is sufficient to account for the radiative decay of the excited state by introducing an imaginary part of energy,

$$E_\varphi \rightarrow \widetilde{E}_\varphi = E_\varphi - i\gamma_\varphi^{\mathrm{rad}}/2, \tag{1.2}$$

so that

$$\varphi(t) = |\varphi\rangle e^{-i\widetilde{E}_\varphi t/\hbar} = |\varphi\rangle e^{-iE_\varphi t/\hbar} e^{-\gamma_\varphi^{\mathrm{rad}} t/2\hbar}. \tag{1.3}$$

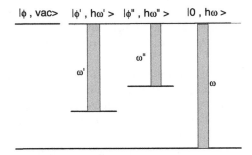

Fig. 1.1. Schematic representation of the excited level φ coupled to a set of radiative continua involving lower φ' levels

The *quasi-stationary state* thus obtained has a survival probability at time t,

$$|\varphi(t)|^2 = e^{-\gamma_\varphi^{\text{rad}} t/\hbar}, \tag{1.4}$$

and decays by fluorescence emission with the *radiative rate*, k_φ^{rad}, related to the radiative lifetime, $\tau_\varphi^{\text{rad}}$, and to the radiative width, $\gamma_\varphi^{\text{rad}}$, by the following relation:

$$k_\varphi^{\text{rad}} = 1/\tau_\varphi^{\text{rad}} = \gamma_\varphi^{\text{rad}}/\hbar. \tag{1.5}$$

In view of the additivity of radiative transition rates, k_φ^{rad} is the sum of contributions of transition rates to all lower lying $|\varphi'\rangle$ states,

$$k_\varphi^{\text{rad}} = \sum_{\varphi'} k_{\varphi\varphi'}^{\text{rad}} = \sum_{\varphi'} A_{\varphi \to \varphi'},$$

where A are familiar Einstein coefficients for spontaneous emission,

$$A_{\varphi \to \varphi'} = \frac{32\pi}{3\hbar} \omega^3 \mu_{\varphi\varphi'}^2,$$

proportional to the square of the transition moment and to the cube of the transition frequency.

In the exact quantal treatment, the radiative continuum contained in the finite volume may be described as a set of closely spaced levels with the level density ρ per cm^{-1}. The decay of the $|\varphi\rangle$ state is induced by its coupling V to this continuum and its decay rate is given by the *Fermi Golden Rule* (cf. [16, 17]),

$$k_{\varphi \to \varphi'} = \frac{2\pi}{\hbar} V^2 \rho. \tag{1.6}$$

This formula is widely applied for description of radiative and nonradiative processes involving the coupling of discrete levels to the continua.

On the other hand, γ^{rad} appears in the light-absorption measurements as the width of the spectral line of the $|0\rangle \to |\varphi\rangle$ transition. The transition induced by a mono-chromatic radiation field involves the linear coupling of the electric field, $\mathbf{F}(\omega, t) = \mathbf{F}_0 e^{i\omega t}$, with the dipole moment of the molecule and is proportional to the square of the interaction matrix element averaged over a time interval larger than ω^{-1},

$$V = \frac{1}{\Delta t} \int_0^{\Delta t} \langle \varphi(t) | \mu \mathbf{F}(t) | 0 \rangle dt. \tag{1.7}$$

The result is

$$|V(\omega)|^2 = \frac{|\mu_{0\varphi}|^2 \mathbf{F}_0^2}{(\omega_{0\varphi} - \omega)^2 + \gamma_\varphi^2/4\hbar^2}, \tag{1.8}$$

which represents a Lorentzian line with a homogeneous width of γ_φ [16].

1.2.2 Zero-Order and Exact Molecular States

The molecular states usually considered in molecular spectroscopy are not the eigenstates, $|\varphi\rangle$, of the exact Hamiltonian, but the eigenstates, $|n\rangle$, of the zero-order Hamiltonian neglecting the spin-orbit coupling and issued from the Born–Oppenheimer approximation. Following the adiabatic separation of the fast electronic movement from slower vibrations and rotations, the state vectors have the forms of ro-vibronic products,

$$|e, v, JKM\rangle = |\psi_e(\overline{Q})\rangle|\chi_{ev}\rangle|JKM\rangle, \qquad (1.9)$$

where the electronic factor, $|\psi_e\rangle$, is defined for a fixed nuclear configuration $\overline{Q}$, the vibrational factor, χ_{ev}, represents the vibrations in the $|\psi_e\rangle$ electronic state and $|JKM\rangle$ characterizes a rigid rotator. In the absence of the spin-orbit coupling, the $|n\rangle$ states are pure spin states, $|\sigma\rangle$ (singlet S_i, and triplet, T_i, states in molecules with an even number of electrons).

We will refer to the basis of the vectors, $|e, v, JKM\rangle|\sigma\rangle$, as to the *zero-order* (ZO) basis. Its interest in spectroscopy is twofold:

- Their energies are determined by simple formulae with a limited number of parameters so that the positions of the whole set of ZO-states with a regular spacing may be easily calculated. The shift and splitting of the observed level with respect to its calculated position are a measure of the coupling omitted in the zero-order approximation,
- The transitions between ZO states are subject to strict selection rules.

We assume that the molecules are vibrationally and rotationally cold ($T_{\text{vib}} \rightarrow 0$ and $T_{\text{rot}} \rightarrow 0$) so that only the lowest rotational levels of the vibrationless electronic state, $|0\rangle$, are populated so that the $|n\rangle$ state may be excited uniquely by the $|0\rangle \rightarrow |n\rangle$ radiative transition. The $|n\rangle$ states are thus divided into two groups:

1. *the bright states*, $|s\rangle$, for which the $|0\rangle \leftrightarrow |s\rangle$ transitions are allowed: ($\mu_{0s} \neq 0$) and
2. *the dark states*, $|\ell\rangle$, for which they are forbidden: ($\mu_{0\ell} = 0$)

In real systems, the $|0\rangle \leftrightarrow |\ell\rangle$ transitions may be not strictly forbidden but so weak that one can consider that $\mu_{0\ell} \approx 0$, as in the case of the $\Delta v \gg 1$ transitions between vibrational levels of the same electronic state. The bright and dark states are sometimes called radiant and nonradiant. In this book, these terms will have a different meaning: dark $|\ell\rangle$ states are *radiant* when radiative $\ell \rightarrow \ell'$ transitions to some lower lying ℓ' states (other than the $|0\rangle$ state) are allowed: $\mu_{\ell\ell'} \neq 0$ so that $\gamma_\ell^{\text{rad}} \neq 0$, whereas for *nonradiant* states, all $\mu_{\ell\ell'} = 0$ and $\gamma_\ell^{\text{rad}} = 0$ (Fig. 1.2). Most of the dark levels are radiant. For instance, the vibronic levels of higher triplet states $|T_{i>1}\rangle$ are dark because the $S_0 \leftrightarrow T_i$ transition is forbidden but are radiant when the $T_i \rightarrow T_{i-1}$ transition to a lower triplet is allowed.

The *exact molecular states* $|\varphi\rangle$ are eigenstates of the complete molecular Hamiltonian. In the zero-order representation, the molecular Hamiltonian has the form $H = H_0 + V$, where H_0 provides the ro-vibronic basis and V accounts for all the

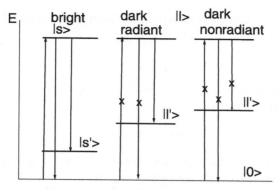

Fig. 1.2. Bright and dark (radiant and nonradiant) states (the forbidden transitions are marked by x)

residual interactions neglected in H_0: the nonadiabaticity (non-Born–Oppenheimer terms), spin-orbit coupling, anharmonicity, Coriolis coupling, etc. which shift and mix the zero-order states. The Hamiltonian matrix is thus written as

$$\begin{matrix} E_n & V_{nn'} & V_{nn''} & \ldots \\ V_{n'n} & E_{n'} & V_{n'n''} & \ldots \\ V_{n''n} & V_{n''n'} & E_{n''} & \ldots \\ \ldots & \ldots & \ldots & \ldots \end{matrix}$$

In the zero-order representation, the eigenstates of the exact molecular Hamiltonian,

$$H|\varphi\rangle = E_\varphi|\varphi\rangle,$$

are given by linear superposition,

$$|\varphi\rangle = \sum_n \alpha_{\varphi n}|n\rangle = \sum_n \langle n|\varphi\rangle|n\rangle, \tag{1.10}$$

whereas

$$|n\rangle = \sum_\varphi \alpha_{n\varphi}|n\rangle = \sum_\varphi \langle\varphi|n\rangle|\varphi\rangle, \tag{1.11}$$

where $\alpha_{\varphi n}$ are elements of the the inverse matrix

$$\alpha_{\varphi n} = \alpha_{n\varphi}^*.$$

The eigenstates $|\varphi\rangle$ are quasi-stationary, however, the zero-order states are not, so their temporal behaviour must be described in terms of exact states

$$|n(t)\rangle = \sum_\varphi \alpha_{n\varphi}|\varphi\rangle e^{-i\widetilde{E}_\varphi t}. \tag{1.12}$$

Because the molecular eigenstates are combinations of bright and dark ZO states,

$$|\varphi\rangle = \alpha_{\varphi s}|s\rangle + \sum_{\ell} \beta_{\varphi \ell}|\ell\rangle, \tag{1.13}$$

the strict selection rules are attenuated. As long as $\alpha_{\varphi s} \neq 0$, the $0 \to \varphi$ transition moment is different from zero,

$$\mu_{0\varphi} = \alpha_{\varphi s}\mu_{0s}, \tag{1.14}$$

whereas all $\mu_{0\ell} = 0$. For instance, the weak singlet-triplet mixing induced by the spin-orbit interaction gives rise to the essentially singlet $|S'\rangle = |S\rangle + \varepsilon|T\rangle$ and essentially triplet $|T'\rangle = |T\rangle - \varepsilon|S\rangle$ states, where $\varepsilon \ll 1$. The $|S_0 \to |T'\rangle$ transition is no more strictly forbidden because $\mu_{S_0 T'} = \varepsilon\mu_{S_0 S}$.

The radiative character of an eigenstate $|\varphi(t)\rangle$ is represented by its projection on the bright state, $\hat{S}(t)$, called S-character,

$$\hat{S}(t) = |\langle s|\varphi(t)\rangle|^2 = |\langle s|\varphi\rangle|^2 e^{-\gamma_\varphi^{rad} t}. \tag{1.15}$$

For instance, the intensity of $|s\rangle \to |s'\rangle$ emission is given by

$$I_s(t) = k_s^{rad}\hat{S}(t), \tag{1.16}$$

where k_s^{rad} denotes the emission rate constant of the bright ZO state. The time dependence of the $\ell \to \ell'$ emission (when the ℓ state is radiant) and of the transient $\ell + \hbar\omega = \ell''$ absorption may be expressed in terms of an analogue, $\hat{L}(t)$ projection.

1.3 Coupling Schemes and Level Patterns

In view of the large variety of molecular systems, the treatment of the coupling schemes must be adapted to different classes of molecules. Their classification is not unique. We consider as the essential criterion the existence or nonexistence of the *continua of energy* (other than the radiative continuum) coupled to discrete energy levels. Such continua may be intrinsic states of the molecule (corresponding, e.g., to dissociation or ionization) or result from the environment effects (continuum of translational energy in the gas or in the fluid solution). As will be discussed in Sect. 1.4.3, the dense sets of molecular levels may be approximated by the *quasi-continua*.

We define as a *small molecule* a system composed of discrete energy levels with widths due uniquely to the coupling to the radiative field (purely radiative widths). The excited state of this system decays radiatively with the overall ($s \to s'$ and $\ell \to \ell'$) emission quantum yield, $Q_f = 1$. On the other hand, in a *large molecule*, discrete levels are coupled to a continuum or quasi-continuum and this coupling implies the broadening (the *nonradiative width*) of discrete levels due to an irreversible transition discrete-to-continuum. In view of a competition between radiative and nonradiative decay channels, the emission yield is reduced, $Q_f \leq 1$. We will apply the term of the *large-molecule limit* to the systems, which may be described in terms of a direct coupling of a single bright state to the continuum (quasi-continuum). We prefer this term to a somewhat misleading term of the *statistical limit*.

The main parameter of each coupling scheme is the average spacing of n-levels, $\langle \Delta E_{nn'} \rangle$, as compared with the average coupling strength $\langle V_{nn'} \rangle$. In the small-molecule limit, the perturbations involving a limited number of levels may be treated exactly in terms of $\Delta E_{s\ell}$ (of the order of $\langle \Delta E_{nn'} \rangle$) and $V_{s\ell}$ values of *individual states*. In view of the increasing number of interacting levels, the coupling in the large molecule limit is described in terms of average *effective level densities*, $\rho_{eff} = 1/\langle \Delta E_{\ell\ell'} \rangle$, and average $\langle V_{s\ell} \rangle$ values.

For a major part of molecular systems, small-molecule and large-molecule schemes do not account for the coupling between discrete $|n\rangle$ ($|s\rangle$ and $|\ell\rangle$) levels coupled in turn to the continuum (or quasi-continuum). We will use, for these systems, the term *sequential coupling case* as more precise than that of the *intermediate coupling case*.

Note that the terms small and large molecule are not necessarily related to their size. So, the predissociation of a diatomic molecule resulting from the coupling to a dissociative continuum must be treated in terms of the large-molecule limit. On the other hand, the lowest vibrational levels of the S_1 state of a molecule as large as 9,10-dichloroanthracene decay by the resonant fluorescence with $Q_f \approx 1$, the coupling to other levels of the S_1 state and to those of T_1 and S_0 states being negligible.

1.3.1 Small Molecules

The interaction between a pair of ZO states $|n\rangle$ and $|n'\rangle$ is significant only in the case of *quasi-resonance*,

$$\frac{|V_{nn'}|}{\Delta E_{nn'}} \gg 0,$$

i.e., when the value of the coupling matrix element $V_{nn'}$ is not much smaller than the energy gap, $\Delta E_{nn'} = |E_n - E_{n'}|$. In a major part of molecular systems, the average spacing between bright $|s\rangle$ levels, $\Delta E_{ss'}$, is so large that one can consider the system as composed of single bright states interacting with one or more dark states. The radiative properties of $|\varphi\rangle$ eigenstates issued from this coupling are

$$\mu_{0\varphi} = \alpha_{\varphi s}\mu_{0s} \qquad \gamma_\varphi^{rad} = |\alpha_{\varphi s}|^2 \gamma_s^{rad} + \sum_\ell |\beta_{\varphi\ell}|^2 \gamma_\ell^{rad} \qquad (1.17)$$

because $\mu_{0\ell} = 0$ for all dark levels, but some of them are radiant ($\gamma_\ell \neq 0$).

The resulting level pattern depends on the relation between average value of the coupling constants $\langle V_{s\ell} \rangle$ and the average ℓ-level spacing $\langle \Delta E_{\ell\ell} \rangle$ expressed usually in terms of the average level density, $\langle \rho_\ell \rangle = 1/\langle \Delta E_{\ell\ell} \rangle$. When $\langle V_{s\ell} \rangle \langle \rho_\ell \rangle \ll 1$, the probability of an s-ℓ quasi-resonance is small so that only a two-level s-ℓ interaction occurs for a fraction of s-levels. This is called the *weak-coupling limit*. In contrast, the *strong-coupling case* corresponds to $\langle V_{s\ell} \rangle \langle \rho_\ell \rangle \geq 1$, where all (or nearly all) s-levels interact with one or more than one ℓ-level.

Weak-Coupling Limit

A good example of the weak coupling is a perturbation between two vibronic states, v_s and v_ℓ, of the s and ℓ electronic states of a diatomic molecule. In view of the conservation of the angular momentum, only the pairs of states with $J_s = J_\ell$ may interact. When the rotational constants in two states are different, $B_s \neq B_\ell$, the $J_s - J_\ell$ level spacing is J-dependent,

$$\Delta E_{s\ell}(J) = \Delta E_{s\ell}(J = 0) + (B_s - B_\ell)J(J + 1) + \cdots, \qquad (1.18)$$

with a quasi-resonance for $J \approx \sqrt{\Delta E_{s\ell}(J = 0)/(B_s - B_\ell)}$. Far from the resonance, $V_{s\ell} \ll \Delta E_{s\ell}(J)$ and the s-ℓ interaction may be neglected. The $|\varphi\rangle$ eigenstates are practically identical with zero-order $|n\rangle$ states, $|\varphi_1\rangle \approx |s\rangle$ bright state and $|\varphi_2\rangle \approx |\ell\rangle$ dark state, so that only the first one appears in the absorption spectrum. In the vicinity of the resonance, the energies E_i and wave functions φ_i ($i = 1, 2$) of eigenstates are given by current equations of a two-level system (cf. [17]) (Fig. 1.3),

$$E_1, E_2 = (E_s + E_\ell)/2 \pm \Delta E/2 \qquad (1.19)$$

where

$$\Delta E_{12} = \sqrt{\Delta E_{s\ell}^2 + 4V_{s\ell}^2} \qquad (1.20)$$

and

$$|\varphi_1\rangle = \cos\theta|s\rangle + \sin\theta|\ell\rangle \quad |\varphi_2\rangle = -\sin\theta|s\rangle + \cos\theta|\ell\rangle, \qquad (1.21)$$

where

$$2\theta = \arctan(2V_{s\ell}/\Delta E_{s\ell}), \qquad 0 \le \theta \le \pi/2, \qquad (1.22)$$

with transition moments

$$\mu_{01} = \cos\theta\mu_{0s}, \quad \mu_{02} = -\sin\theta\mu_{0s} \qquad (1.23)$$

and widths

$$\gamma_1 = \cos^2\theta\gamma_s + \sin^2\theta\gamma_\ell, \quad \gamma_2 = \sin^2\theta\gamma_s + \cos^2\theta\gamma_\ell. \qquad (1.24)$$

Note that, by definition, $\theta \le \pi/4$, so that $\cos^2\theta \ge \sin^2\theta$.

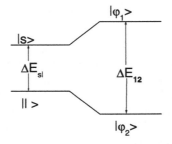

Fig. 1.3. A scheme of the two-level coupling

The single absorption line corresponding to the $|0\rangle \rightarrow |s\rangle$ transition (the $|0\rangle \rightarrow |\ell\rangle$ transition being forbidden in the absence of perturbation) is replaced for perturbed levels by two lines with intensities $I_\varphi = I_1 = I_s \cos^2 \theta$ and $I_{\varphi'} = I_2 = I_s \sin^2 \theta$, the weaker one being usually called extra-line. We will use in the following the notation $|\varphi\rangle$ and $|\varphi'\rangle$ for eigenstates with a large and small s-content.

In the weak perturbation limit ($V_{s\ell}^2 \ll \Delta E_{s\ell}$), Eq. 1.20 may be rewritten as

$$\Delta E_{12} = \frac{1}{2}\Delta E_{s\ell}\sqrt{(1 + V_{s\ell}^2/\Delta E_{s\ell}^2)^2} \approx \Delta E_{s\ell} + V_{s\ell}^2/\Delta E_{s\ell}; \qquad (1.25)$$

hence, $E_1 = E_s + V_{s\ell}^2/\Delta E_{s\ell}$ and $E_2 = E_\ell - V_{s\ell}^2/\Delta E_{s\ell}$.

In this treatment, the level widths (imaginary parts of energy) are neglected, assuming that $\gamma_s - \gamma_\ell \approx 0$, so that $\widetilde{\Delta E_{s\ell}} = \Delta E_{s\ell}$.

The rigorous treatment of the coupling including level widths is complicated (cf. [18]), but in the limiting case of a weak perturbation, we have

$$\Delta E_{12} = \Delta \widetilde{E_{s\ell}} + \frac{1}{2}\frac{V_{s\ell}^2}{\Delta E_{s\ell} - i\Delta\gamma_{s\ell}/2},$$

so that

$$E_1 = E_s + \frac{\Delta E_{s\ell} V_{s\ell}^2}{\Delta E_{s\ell}^2 + \Delta\gamma_{s\ell}^2/4}, \qquad \gamma_1 = \gamma_s - \frac{\Delta\gamma_{s\ell} V_{s\ell}^2/2}{\Delta E_{s\ell}^2 + \Delta\gamma_{s\ell}^2/4}$$

$$E_2 = E_\ell - \frac{\Delta E_{s\ell} V_{s\ell}^2}{\Delta E_{s\ell}^2 + \Delta\gamma_{s\ell}^2/4}, \qquad \gamma_2 = \gamma_\ell + \frac{\Delta\gamma_{s\ell} V_{s\ell}^2/2}{\Delta E_{s\ell}^2 + \Delta\gamma_{s\ell}^2/4}.$$

The main implication of difference in level widths is the attenuation of the s-ℓ interaction (s-ℓ decoupling), important when the $\Delta\gamma$ term in denominators is non-negligible as compared with ΔE.

The Strong Coupling Case

The coupling of one s-state to several ℓ-states gives rise to a band of $|\varphi\rangle$ states with wave functions given by Eq. 1.13. The $\alpha_{\varphi s}$ and $\beta_{\varphi\ell}$ coefficients, radiative properties and energies E_φ energies of resulting states may be calculated by diagonalization of the $\|V_{s\ell}\|$ matrix if energies and coupling constants of relevant ZO states are known. This information is usually nonavailable and only rough estimates of the φ-level pattern may be made by assuming either a regular distribution of ℓ-levels: equidistant and equally coupled to $|s\rangle$ or a random variation of $\Delta E_{s\ell}$ and $V_{s\ell}$ parameters. In the first case, the picture is that of a band of N nearly equidistant lines with a quasi-Lorentzian intensity distribution and an overall width, Γ (Fig. 1.4 (a)), where

$$\Gamma = 2\pi V_{s\ell}^2 \rho_\ell \qquad (1.26)$$

$$N = \Gamma\rho_\ell = 2\pi V_{s\ell}^2 \rho_\ell^2. \qquad (1.27)$$

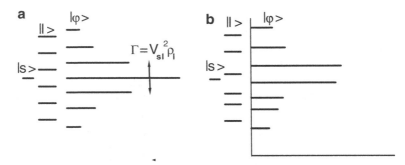

Fig. 1.4. Scheme of the strong coupling of (a) equidistant and (b) randomly distributed ℓ-levels

On the other hand, if a random distribution of $\Delta E_{s\ell}$-level spacings and $V_{s\ell}$ coupling strength around their average values is assumed, the frequencies and intensities of lines vary in wide limits but the overall width of the band is of the order of magnitude of (Fig. 1.4 (b))

$$\Gamma \approx 2\pi \langle V_{s\ell}^2 \rangle \rho_\ell, \tag{1.28}$$

basically the same as that predicted by the model of regular states.

The energies E_s and E_ℓ of zero-order states as well as their coupling constants, $V_{s\ell}$, cannot be directly measured. They are deduced from observed frequencies and intensities of $|0\rangle \rightarrow |\varphi\rangle$ transitions by the deconvolution technique [19].

1.3.2 Large Molecule Limit

We define as a large molecule a molecular system in which a single discrete level, $|s\rangle$, is directly coupled to a continuum (or to a dense level set approximated by a continuum) of $\{m\}$ states.

The best example of a large molecule is, paradoxically, the electronic predissociation of a diatomic XY molecule. A discrete bright XY* level with energy above the XY* $\rightarrow$ X + Y dissociation threshold is directly coupled to the continuum of recoil energies of a free X + Y pair and to the radiative (XY + $\hbar\omega$) continuum. The two channels do not interfere so that the level width is a simple sum of the radiative γ_s^{rad} and nonradiative γ_s^{nr} widths,

$$\gamma_s = \gamma_s^{\text{rad}} + \gamma_s^{\text{nr}}. \tag{1.29}$$

Because the s-m coupling strength is the same for the whole set of identical $\{m\}$ levels, the nonradiative width of $|s\rangle$ as well as its radiative width is given by the Fermi Golden Rule,

$$\gamma_s^{\text{nr}} = 2\pi V_{sm}^2 \rho_m, \tag{1.30}$$

where ρ_m is the density of m-levels. The shape of a $|0\rangle \rightarrow |s\rangle$ absorption line is Lorentzian.

The processes involving the energy transfer from the initially excited discrete level of a small molecule to the continuum of the kinetic energy (collisional relaxation, fluorescence quenching) may be treated, at least in the first approximation, in terms of the large-molecule limit.

In a large polyatomic molecule, the vibrationless level $|s\rangle$ of the excited electronic state is coupled to a dense set $\{m\}$ of high vibrational levels of lower electronic states. These levels do not form a true continuum in an isolated molecule, but they are so strongly mixed that they may be approximated by a quasi-continuum of levels, nearly equidistant and equally coupled to $|s\rangle$ and well characterized by the average values of the coupling constant $\langle V_{sm}\rangle$ and of the level density ρ_m. The nonradiative level width is then given by the Fermi Golden Rule in the form

$$\gamma_s^{nr} \approx 2\pi|\langle V_{sm}\rangle|^2\rho_m. \tag{1.31}$$

This picture implies an irreversible decay of the s-level. The necessary conditions of the irreversibility will be discussed in Sect. 1.4.3.

1.3.3 Sequential-Coupling Model

The essential parameter of this model is the relation between the strengths of the s-ℓ, s-m and ℓ-m interactions. The system of a particular interest is the system in which the s-level is selectively coupled to a small number of dark ℓ levels coupled in turn to the m manifold, whereas the direct s-m coupling is weak. As in the small-molecule case, the s-ℓ coupling gives rise to a set of discrete φ-states with nonradiative widths resulting from the ℓ-m coupling,

$$\gamma_\varphi = |\alpha_{\varphi s}|^2\gamma_s + \sum_\ell |\beta_{\varphi\ell}|^2\gamma_\ell + \gamma_\ell^{nr}, \tag{1.32}$$

where

$$\gamma_\ell^{nr} = 2\pi|\langle V_{\ell m}\rangle|^2\rho_m.$$

The structure of the absorption spectrum depends on the relation between the spacings $\Delta E_{\varphi\varphi'}$ and widths γ_φ of the $|\varphi\rangle$ states. When the ℓ-m coupling is weak, so that $\gamma_\ell^{nr} \to 0$ and

$$\gamma_\varphi \ll \Delta E_{\varphi\varphi'},$$

the individual $|0\rangle \to |\varphi\rangle$ transitions are well resolved. In contrast, when

$$\gamma_\ell^{nr} \geq \Delta E_{\varphi\varphi'},$$

the homogeneously broadened $|0\rangle \to |\varphi\rangle$ transitions overlap so that the structure of the absorption band is partially or entirely washed out. When the widths of individual $|\varphi\rangle$ states become of the same order of magnitude as the Γ width of the whole band of φ-states,

$$2\pi V_{\ell m}^2\rho_m \approx \Gamma,$$

the broad and strongly overlapping lines coalesce into a single structureless absorption band, as in the large-molecule limit. We will, however, show (Sec. 5.2) that, even in this case, the properties of the excited state are different from the large-molecule behaviour.

1.4 The Time Evolution of Excited States

When the duration of the exciting light pulse δt_{rad} is very short as compared with the rate of the further evolution of the excited system, the whole process may be divided into preparation of the initial state, $\Psi(0)$, and its further evolution $\Psi(t)$ in the absence of the external radiation field. We will consider the most elegant and the simplest case of excitation by the light pulse issued from a coherent light source, such a single-mode or mode-locked laser producing strictly identical radiation wave packets with Fourier-limited coherent width and duration.

The light pulse interacts with the molecular system via its electric field,

$$F(\omega, t) = \int_0^\infty F_0(\omega)e^{i\omega t}\, d\omega,$$

where the frequency distribution function $F_0(\omega)$ is centered at ω_0 with the width $\delta\omega_{coh}$ and whereas the pulse duration is $\delta t = \delta\omega_{coh}^{-1}$. The linear coupling of the field to the molecular dipole moment creates a coherent superposition of molecular states,

$$|\Psi(0)\rangle = \mu F|0\rangle = \sum_\varphi F_0(\omega_{0\varphi})\mu_{0\varphi}|\varphi\rangle, \tag{1.33}$$

where $\omega_{0\varphi}$ and $\mu_{0\varphi}$ represent, respectively, the transition frequency and transition moment.

The molecular state prepared in this way may be represented by a vibrational wave packet, $\Psi(Q, t)$. The interest of this description derives in part from the fact that it provides the possibility of a unified treatment of bound and free states and of their superposition, e.g., in the case of the excitation of levels in the vicinity of a dissociation threshold.

1.4.1 Vibrational Wave Packet

We consider initially the time-dependent wave function for a dissociation continuum (free particle), which we write as

$$\Psi(Q, t) = \int dk\, g(k)e^{i[kQ-\omega(k)t]}, \tag{1.34}$$

in terms of progressive waves characterized by a local wave number k related to ω by

$$\omega(k, Q) = \frac{E - V(Q)}{\hbar} \equiv \frac{\hbar k^2(Q)}{2m}. \tag{1.35}$$

Here, $g(k)$ is assumed to be real with a distribution peaked at $k = k_0$ and a width $k_0 - \Delta k/2 \ldots k_0 + \Delta k/2$. We expand $\omega(k)$ near $k = k_0$ (where the constructive interference occurs) as

$$\omega(k) \approx \omega(k_0) + (k - k_0)\left(\frac{\partial\omega}{\partial k}\right)_{k=k_0}, \tag{1.36}$$

which, when substituted into Eq. 1.34, yields

$$\Psi(Q, t) \approx e^{i[k_0 Q - \omega(k_0)t]} \int dk \, g(k) e^{i(k-k_0)\left(Q - (\partial\omega/\partial k)_{k=k_0} t\right)}. \tag{1.37}$$

From Eq. 1.35, we find that

$$\frac{\partial\omega}{\partial k} = \frac{\hbar k}{m} = \frac{p_{cl}}{m} = v_{cl}, \tag{1.38}$$

where p_{cl} is the classical momentum and v_{cl} is the classical velocity. Eq. 1.37 thus becomes

$$\Psi(Q, t) \approx e^{i[k_0 Q - \omega(k_0)t]} \int dk \, g(k) e^{i(k-k_0)(Q - v_{cl}t)}. \tag{1.39}$$

Constructive interference takes place for $k = k_0$ (where $g(k)$ has its maximum) when the phase in Eq. 1.39 is (nearly) independent of the wave number k, i.e., when $Q \approx v_{cl}t$. This is the so-called *stationary phase condition*. It shows that the peak of the wave packet at time t occurs near $Q = v_{cl}t$ as expected on the basis of classical physics.

The width of the packet at $t = 0$ is given according to the uncertainty relation by $\Delta Q(t = 0) \approx \hbar/\Delta p_{cl} = 1/\Delta k$. Indeed, we see from Eq. 1.34 that the wave packet goes to zero when destructive interference occurs. This happens when the components with $k_0 \pm \Delta k/2$ are out of phase by $\pm\pi$, which is the case for $Q \approx \pm 2\pi/\Delta k$. The spreading (or bunching) of the packet occurs because the classical velocities of its components differ slightly, and it obviously depends on the shape of the potential $V(Q)$. Consider a constant potential, $V(Q) = $ const, and focus on the point $Q = Q_0$, where the center of the packet arrives at $t_0 = Q_0/v_{cl}$. The component $k_0 - \Delta k/2$ arrives with a delay $\Delta t = (mQ_0/\hbar k_0)(\Delta k/k_0)$ after the component corresponding to $k_0 + \Delta k/2$. Because the group velocity of the packet is v_{cl}, this delay implies a spread of the packet $\Delta Q(t) = v_{cl}\Delta t$. We thus find that

$$\Delta Q(t) \approx \frac{\hbar}{m} \Delta k \, t. \tag{1.40}$$

This expression neglects the initial spread of the wave packet ($t = 0$) and is therefore valid only for large times, $t \gg (m/\hbar)(1/[\Delta k]^2)$, for which it shows the spread of a free wave packet that proceeds with constant velocity, to increase linearly with time. We thus see that the free wave packet is initially ($t = 0$) the sharper the *larger* Δk, whereas by contrast, at long times, it spreads more slowly the *smaller* Δk.

In the case of bound vibrational wave packets, the integration over k in Eq. 1.34 must be replaced by a sum over bound eigenstates of the vibrational Hamiltonian. For example, a one-dimensional vibrational bound wave packet (solution of the time-dependent Schrödinger equation) expanded in terms of the solutions of the time-independent Schrödinger equation is, in analogy with Eq. 1.34,

$$\Psi(Q, t) = \sum_v c_v \chi_v(Q) e^{-i[(E_v/\hbar)t + \delta_v]}, \tag{1.41}$$

where $\chi_v(Q)$ are the eigenfunctions and E_v the eigenenergies of the time-independent vibrational Hamiltonian and the coefficients c_v and phases δ_v depend on the initial conditions of excitation of the wave packet. If the vibrational states are those of a harmonic oscillator with $E_v = \omega_e(v + 1/2)$, we see from Eq. 1.41 that $\Psi(Q, t)$ is strictly periodic with period $T = 4\pi\hbar/\omega_e$, i.e., there is no dispersion or broadening of the packet in this case.

1.4.2 Coherent and Incoherent Excitation

We are interested in a particular case of interaction between the wave packet of radiation with the coherent width $\delta\omega_{\mathrm{coh}}$ and the band of φ states issued from the coupling of a single bright $|s\rangle$ state with several $|\ell\rangle$ states with the average spacing $\delta E_{\varphi\varphi'}$ and the overall width Γ.

The coherent excitation of the whole band of states takes place when $\delta\omega_{\mathrm{coh}} > \Gamma$. The excited state is described by Eq. 1.33. If assumed that the amplitude F_0 is approximately constant over the whole frequency range,

$$|\Psi(0)\rangle = \sum_\varphi \mu_{0\varphi}|\varphi\rangle \qquad (1.42)$$

because $\mu_{0\varphi} = \alpha_{\varphi s}\mu_{0s}$, this superposition is identical with the bright state,

$$|s\rangle \sim \mu|0\rangle = \sum_\varphi \alpha_{\varphi s}|\varphi\rangle. \qquad (1.43)$$

The opposite limiting case is a selective excitation of a single molecular eigenstate, which requires a pulse with $\delta\omega_{\mathrm{coh}} \ll \Delta E_{\varphi\varphi'}$ and $\omega_0 = \omega_{0\varphi}$. The initially excited state is then simply $\Psi(0) = |\varphi\rangle$.

The incoherent excitation of several molecular eigenstates is a rule in the case of *chaotic light sources* such as a multimode lasers or an electric discharge. Their radiation field may be regarded as an incoherent mixture of monochromatic components distributed around a central frequency with the overall width $\delta\omega_{\mathrm{tot}}$; the excited system is then an incoherent mixture of molecular eigenstates. When $\omega_{\mathrm{tot}} > \Gamma$, the whole band of $|\varphi\rangle$ states is excited, but each of them decays independently with its characteristic rate $\gamma_\varphi/\hbar$.

1.4.3 The Time Evolution of the Excited Molecule

We will treat separately three types of molecular systems: that composed of discrete s and ℓ energy levels (small molecule), a discrete s level directly coupled to the dissipative continuum $\{m\}$ (large-molecule limit) and more complex systems with a sequential s-ℓ-$\{m\}$ coupling.

Small Molecule

In the absence of dissipative $\{m\}$ continua, all molecular states decay radiatively with the rate $k_\varphi = k_\varphi^{\text{rad}} = \gamma_\varphi^{\text{rad}}/\hbar$ and the overall emission quantum yield $Q_f = Q_f^s + Q_f^\ell = 1$. If the $s \to s'$ and $\ell \to \ell'$ emission components are spectrally separable, their yields and the time dependence of emission intensities $I_s(t)$ and $I_\ell(t)$ may be measured separately.

There is no fundamental difference between the many-level $s - \ell$ quasi-resonances (the strong-coupling case) and one-to-one s-ℓ interaction in the weak coupling limit. We will treat the time evolution in terms of previously defined $\hat{S}(t)$ and $\hat{L}(t)$ projections of $\Psi(t)$ on the $|s\rangle$ and $|\ell\rangle$ state manifold.

Incoherent Excitation

Upon the excitation of a single φ level,

$$|\varphi\rangle = \alpha_{\varphi s}|s\rangle + \sum_\ell \beta_{\varphi\ell}|\ell\rangle,$$

we have

$$\hat{S}(t) = |\alpha_{\varphi s}|^2 e^{-\gamma_\varphi t/\hbar} \tag{1.44}$$

and

$$\hat{L}(t) = \sum_\ell |\beta_{\varphi\ell}|^2 e^{-\gamma_\varphi t/\hbar} = (1 - |\alpha_{\varphi s}|^2) e^{-\gamma_\varphi t/\hbar}, \tag{1.45}$$

so that $I_s(t) = k_s\hat{S}(t)$ and $I_\ell(t) = k_\ell\hat{L}(t)$ decay exponentially with the same rate k_φ,

$$k_\varphi = |\alpha|_{\varphi s}^2 k_s + \sum_\ell |\beta|_{\varphi\ell}^2 k_\ell.$$

If the ℓ-states are nonradiant or weakly radiant, $(\gamma_s^{\text{rad}} \gg \gamma_\ell^{\text{rad}})$, the decay rate of the $|\varphi\rangle$ state is reduced as compared with that of the $|s\rangle$ state. This reduction may be roughly estimated if assumed that the s-state is completely diluted in a band of $N - 1$ ℓ-states,

$$|\varphi\rangle = 1/\sqrt{N}(|s\rangle + \sum_\ell |\ell\rangle), \tag{1.46}$$

and that $\gamma_\ell \approx 0$. In this limit, the decay rates of all φ-levels issued from the same bright s-state are identical,

$$\gamma_\varphi \approx \gamma_s/N + \gamma_\ell \approx \gamma_s/N. \tag{1.47}$$

In real systems, there is a wide distribution of the $|\alpha_{\varphi s}|^2$ coefficients and of γ_φ values, but γ_s/N corresponds to the average of this distribution.

Upon an incoherent excitation of a set of φ levels, the time dependence of the $|s\rangle \to |s'\rangle$ emission intensity is a sum of exponentials,

$$I(t) \approx \sum_\varphi |\alpha_{\varphi s}|^2 e^{-|\alpha_{\varphi s}|^2 k_s t}.$$

In the *weak-coupling limit*, the incoherent excitation prepares a mixture of two states, $|\varphi_1\rangle = \cos\theta|s\rangle + \sin\theta|\ell\rangle$ and $|\varphi_2\rangle = -\sin\theta|s\rangle + \cos\theta|\ell\rangle$, with the initial weights proportional to transition probabilities $N_1 \propto \mu_{01}^2 = N\cos^2\theta$ and $N_2 \propto \mu_{02}^2 = N\sin^2\theta$. We can record separately the $s \rightarrow s'$ and $\ell \rightarrow \ell'$ emission components, but the emission from φ_1 and φ_2 close-lying levels cannot be separated. One can thus record

$$I_s(t) = k_s[N_1|\langle s|\varphi_1\rangle|^2 e^{-k_1 t} + N_2|\langle s|\varphi_2\rangle|^2 e^{-k_2 t}]$$
$$= k_s[\cos^4\theta e^{-k_1 t} + \cos^2\theta \sin^2\theta e^{-k_2 t}] \qquad (1.48)$$
$$I_\ell(t) = k_\ell[N_1|\langle \ell|\varphi_1\rangle|^2 e^{-k_1 t} + N_2|\langle \ell|\varphi_2\rangle|^2 e^{-k_2 t}]$$
$$= k_\ell[\cos^2\theta \sin^2\theta e^{-k_1 t} + \sin^4\theta e^{-k_2 t}], \qquad (1.49)$$

where

$$k_1 = k_s\cos^2\theta + k_\ell\sin^2\theta, \quad k_2 = k_s\sin^2\theta + k_\ell\cos^2\theta. \qquad (1.50)$$

Because, by definition, $\cos^2\theta \geq \sin^2\theta$, the decays of both emission components are biexponential: a stronger short-lived component and a weaker long-lived one correspond to light absorbed in the main and in the extra absorption line (Fig. 1.5 (a)). In the case of an exact s-ℓ resonance, $\cos^2\theta = \sin^2\theta = 1/2$ and both φ-states decay with the same $\gamma = (k_s + k_\ell)/2$ rate.

Coherent Excitation

The excitation of the entire set of $|\varphi\rangle$ states sharing the transition moment of an $|s\rangle$ state by a single wave packet with $\delta_{\mathrm{coh}} \gg \Gamma$ prepares a coherent superposition of states with amplitudes proportional to their $\alpha_{\varphi s}$ coefficients, $\Psi(0) = \sum_\varphi \alpha_{s\varphi}|\varphi\rangle$. Because $\sum_\varphi \alpha_{s\varphi}^2 = 1$, this state is identical with the zero-order bright state $|s\rangle$,

$$\langle s|\Psi(0)\rangle = \sum_\varphi \alpha_{s\varphi}\langle s|\varphi\rangle. \qquad (1.51)$$

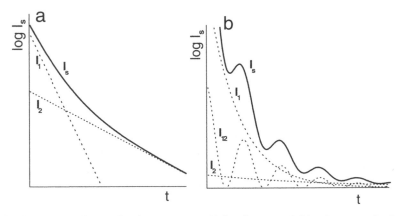

Fig. 1.5. Decay of a two-level system upon (a) incoherent and (b) coherent excitation

So in the weak coupling limit,

$$\Psi(0) = \cos\theta|\varphi_1\rangle - \sin\theta|\varphi_2\rangle = |s\rangle. \tag{1.52}$$

Its time evolution is given by the projections of $\Psi(t)$, expressed on the basis of φ eigenstates, on the $|s\rangle$ and $|\ell\rangle$ states,

$$|\langle s|\Psi(t)\rangle|^2 = \cos^4\theta e^{-k_1 t} + \sin^4\theta e^{-k_2 t}$$
$$+ \cos^2\theta\sin^2\theta\cos^2[(\omega_1 - \omega_2)t]e^{-(k_1+k_2)t} \tag{1.53}$$

$$|\langle\ell|\Psi(t)\rangle|^2 = \sin^4\theta e^{-k_1 t} + \cos^4\theta e^{-k_2 t}$$
$$+ \cos^2\theta\sin^2\theta\sin^2[(\omega_1 - \omega_2)t]e^{-(k_1+k_2)t}, \tag{1.54}$$

where $\omega_1 - \omega_2 = \Delta E_{12}/\hbar$.

The decay curve is thus a sum of two exponentials, $I_1(t)$ and $I_2(t)$, with k_1 and k_2 decay rates, as in the case of incoherent excitation but contains the third term $I_{12}(t)$, representing coherence between two states. This term decays with an average $(k_1 + k_2)/2$ rate and is modulated by the frequency corresponding to the $|E_1 - E_2|$ energy difference (Fig. 1.5 (b)), giving rise to the *two-level quantum beats*. This picture is still simpler in the case of an exact resonance between $|s\rangle$ and $|\ell\rangle$ levels. In this case,

$$k_1 = k_2 = k, \quad \cos\theta = \sin\theta, \quad \hbar|\omega_1 - \omega_2| = 2V_{s\ell},$$

and the decay is monoexponential and entirely modulated by the beats,

$$\hat{S}(t) = \cos^2[(\omega_1 - \omega_2)t]e^{-kt} \quad \hat{L}(t) = \sin^2[(\omega_1 - \omega_2)t]e^{-kt}, \tag{1.55}$$

with a $\pi/2$ phase shift between $\hat{S}(t)$ and $\hat{L}(t)$. The excitation prepares the $|s\rangle$ state so that $\hat{S}(0) = 1$ and $\hat{L}(0) = 0$. Then $\hat{L}$ builds up when $\hat{S}$ decays.

Similar formulae describe the time evolution of the superposition of N coherently excited states, $|\varphi_i\rangle$, resulting from the s-ℓ coupling in the strong-coupling case. The decay of $\hat{S}(t)$ and $\hat{L}(t)$ is composed of N exponentials, $A_\varphi(t)$, with k_φ rates and of $N(N-1)/2$ terms, $B_{\varphi\varphi'}(t)$, describing the beats between the pairs of levels φ and φ',

$$[a_\varphi a_{\varphi'}\cos^2\Delta\omega_{\varphi\varphi'}t + \phi]e^{-(k_\varphi+k_{\varphi'})t/2}, \tag{1.56}$$

where $\phi = 0$ and $\pi/2$ for $\hat{S}(t)$ and $\hat{L}(t)$, respectively.

The shape of decay curves depends on the structure of the level system. In an idealized model of a set of ℓ-states equidistant and equally coupled to $|s\rangle$, the eigenstates may be also approximated by a set of states with identical $\omega_\varphi - \omega_{\varphi'} = \Delta\omega$ level spacing and identical $k_\varphi = k_{\varphi'}$ rates. In such a system, $\hat{S}(t)$ shows initially an exponential decay with the rate $k = \Gamma/\hbar$ corresponding to the loss of the initial constructive interference within the set of $B_{\varphi\varphi'}(t)$ terms [8]. This interference reappears after the recurrence times, $t_{rec} = 2n\pi/\Delta\omega$, so that the decay of $\hat{S}(t)$ is composed of narrow peaks with duration $\delta t \approx \hbar/\Gamma$ separated by dark periods, $\Delta t \approx 1/\Delta\omega$.

Such a regular level structure is unusual in real molecules, but this model is useful for estimation of time scales. In the case of the frequency differences $\omega_\varphi - \omega_{\varphi'}$ varying

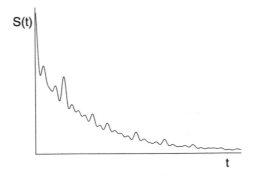

Fig. 1.6. Simulation of the decay of a system of four randomly distributed and coherently excited $|\varphi\rangle$ levels

randomly in the 0 to $\Gamma/\hbar$ limits, the initial phase agreement is also lost at the $\Gamma/\hbar$ time scale. The recurrence times $t_{\varphi\varphi'}$ are different for different pairs of levels so that the initial constructive interference of levels is never recovered. For a small number of coherently excited levels, a quasi-exponential decay is more or less deeply modulated by *many-level quantum beats* (Fig. 1.6). For a larger number of levels, the structure of beats is washed out and only the initial signal drop due to the dephasing is clearly seen. Such a decay may be roughly fitted by a biexponential: a short-lived emission from the initially prepared $|s\rangle$ state, with a decay rate $k_1 \approx \Gamma/\hbar$, and a long-lived fluorescence, with a rate $k_2 \approx (\gamma_s/N + \gamma_\ell)/\hbar$ corresponding to the average decay rate of φ-states.

Large Molecules

Large-Molecule Limit

The coupling of the initially excited $|s\rangle$ state to all other discrete levels of the molecule being neglected, the decay of the $|s\rangle$ state is due uniquely to its coupling to the continua: the radiative continuum and the quasi-continuum of dark m-states. The rate of the $s \rightarrow \{m\}$ irreversible transition is given by the Fermi Golden Rule, $k_s^{nr} = 2\pi V_{sm}^2 \rho_m/\hbar$. The radiative and nonradiative channels, with k_s^{rad} and k_s^{nr} rates, do not interfere, so that $k_s = k_s^{rad} + k_s^{nr}$. The decay of the s-state and the rise of the m-state populations are described by the exponentials with the same constants,

$$\widehat{S}(t) = e^{-k_s t} \quad \widehat{M}(t) = \frac{k_s^{nr}}{k_s}(1 - e^{-k_s t}). \tag{1.57}$$

The lifetime of the s-state fluorescence, $\tau_s = 1/(k_s^{rad} + k_s^{nr})$, is shorter than its radiative lifetime, $\tau_s^{rad} = 1/k_s^{rad}$, and its quantum yield,

$$Q_f^s = \frac{k_s^{rad}}{k_s^{rad} + k_s^{nr}} < 1 \tag{1.58}$$

is reduced as compared with $Q_f = 1$ in small molecules.

From the Strong-Coupling Case to the Large-Molecule Limit

In the strong-coupling case, a bright $|s\rangle$ level is coupled to a set of discrete levels, whereas in the large-molecule limit, this set is considered as a continuum. Because there are no continua in the energy spectrum of an isolated molecule, it is necessary to define the conditions that allow us to approximate a dense level set by a continuum.

We will consider as criterion the relation between the average spacing of the neighbour m-levels, $\langle \Delta E_{mm'} \rangle$, and their widths, γ_m. As previously discussed, the average recurrence time, t_{rec}, is inversely proportional to the level spacing,

$$\langle t_{rec} \rangle \approx \frac{1}{\langle \Delta E_{mm'} \rangle / \hbar}.$$

The recurrences are thus more and more delayed when the level spacing decreases, i.e., when the level density, ρ_m, increases but as long as the widths of dark levels may be neglected ($\gamma_\ell \approx 0$), the evolution of $\hat{S}(t)$ remains periodic and $Q_f = 1$.

The dark levels have, however, nonzero widths, $\gamma_m \neq 0$, due to their radiative and nonradiative intrinsic decay channels, to collisional effects, etc. When the decay time of dark states, $\tau_m = \hbar/\gamma_m$, is shorter than t_{rec}, the recurrences are quenched, the decay of $\hat{S}$ becomes irreversible and the dense set of discrete levels behaves as a dissipative continuum m. The transition from the strong-coupling case to the large-molecule limit takes place when

$$\tau_m < t_{rec} \quad \leftrightarrow \quad \gamma_m > \langle \Delta E_{mm'} \rangle, \tag{1.59}$$

both conditions being strictly equivalent. This transition may be thus induced either by increasing level density (e.g., substitution of the aromatic ring with methyl groups) or by level broadening due to the environment (e.g., collisional) effects.

The Sequential Coupling Case

The strong-coupling as well as large-molecule schemes are based on a "democratic" model of nearly equal coupling of all dark levels to the bright s-level. A closer insight suggests that, in a large fraction of molecular systems, the s-state is efficiently coupled only to a limited number of ℓ-states, some of them being coupled in turn to a dense $\{m\}$-level set, a direct s-m coupling being negligible. As already discussed (Sect. 1.3.3), the properties of the system depend on the relation between the s-ℓ and ℓ-m coupling strength. We will consider here two limits: $V_{s\ell} \gg V_{\ell m}$ and $V_{s\ell} \ll V_{\ell m}$ in the simplest case of the exact resonance between the levels: $|\ell\rangle$ and $|s\rangle$ giving rise to $|1\rangle$ and $|2\rangle$ φ eigenstates split by $\Delta E_{12} = 2V_{s\ell}$. For simplicity sake, we assume that the width of the s-state is purely radiative whereas that of the nonradiant ℓ state is induced by its coupling to the m quasi-continuum,

$$\gamma_s = \gamma_s^{rad}, \quad \gamma_\ell = \gamma_\ell^{nr} = V_{\ell m}^2 \rho_m. \tag{1.60}$$

When the splitting of φ states is large as compared with their widths, the selective excitation of each of them is possible and gives rise to an exponential decay with identical rates, $k_1 = k_2 = (\gamma_s + \gamma_\ell)/2\hbar$, and the fluorescence yield,

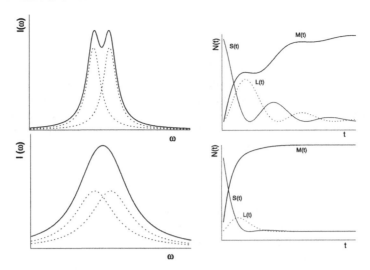

Fig. 1.7. Absorption spectra (left) and time evolution of a two-level system coupled to the continuum for $\gamma / \Delta E = 1/2$ (top) and $\gamma / \Delta E = 3/2$ (bottom)

$$Q_f = \gamma_s / (\gamma_s + \gamma_\ell).$$

Upon coherent excitation, the exponential decay of $\hat{S}(t)$ and $\hat{L}(t)$ is modulated by the two-level quantum beats as in the strong-coupling case. Because the m levels are populated by the $|\ell\rangle \rightarrow m$ transition, $\widehat{M}(t)$ builds up with the rate $d\hat{M}/dt = k_\ell^{nr} \hat{L}(t)$. The rise of $\hat{M}(t)$ is not directly related to the $\hat{S}(t)$ decay as in the large-molecule limit. Moreover, the rising part of the $\hat{M}(t)$ curve shows a stepwise increase, reflecting the beats in the $\hat{L}(t)$ decay (Fig. 1.7 (a)). A similar stepwise rise will occur with the $\pi/2$ delay when m levels are populated from s- and not from the ℓ-state. Such an exotic behaviour was observed in the case of the excited NaI molecule (Sect. 5).

When $V_{\ell m}^2 \rho_m \geq \Delta E_{12}$, the $|\ell\rangle$ level is so strongly broadened that the treatment of the decay in terms of φ states loses its interest. In view of the strong ℓ-m coupling, it is more interesting to diagonalize the $\ell + m$ level system and to treat the coupling of the $|s\rangle$ level to a set of N $\{\mu\}$ states resulting from the ℓ level mixing with $N - 1$ $\{m\}$ states,

$$|\mu\rangle = (1/\sqrt{N}) \left(|\ell\rangle + \sum_{m}^{N-1} |m\rangle \right). \tag{1.61}$$

Because the direct s-m coupling may be neglected, the effective s-μ coupling constant is

$$V_{s\mu} = V_{s\ell}/\sqrt{N}.$$

If assumed that the number N of $\{m\}$ states coupled to ℓ is proportional to the ratio of level densities, i.e., $N \approx \rho_m/\rho_\ell$, the nonradiative decay rate of the s-state,

$$k_s^{nr} = |V_{s\mu}|^2 \rho_\mu \approx (|V_{s\ell}|^2/N) N\rho_\ell = |V_{s\ell}|^2 \rho_\ell, \tag{1.62}$$

is independent of the m-level density. The relaxation rate is thus limited by its slow step, $|s\rangle \rightarrow |\ell\rangle$, and is not sensitive to the rate of the rapid step, $|\ell\rangle \rightarrow m$, which is dependent on the m-level density.

The rise of $\hat{M}(t)$ is delayed with respect to the decay of $\hat{S}(t)$, as can be seen in Fig. 1.7 (b), but this delay is small and may be easily overlooked. In contrast, the *independence of the $\hat{S}(t)$ decay rate of the overall level density* is an important feature, specific for the sequential-coupling case, which differentiates it from the large-molecule limit.

The sequential-coupling model limited to three (s, ℓ and m) level manifolds is still oversimplified and not sufficient to describe in a detailed way the time evolution of a number of molecular systems. A more general *tier model* assuming the sequential coupling between different tiers of states, $s \leftrightarrow \ell_1 \leftrightarrow \ell_2 \leftrightarrow \cdots \leftrightarrow m$ [20], is more efficient provided that the energies and coupling constants are known for the whole set of relevant states.

1.5 Summary

The *small-molecule system* is a system of discrete energy levels uncoupled from the dissipative continua intrinsic or induced by the interaction with the medium (heat bath). The optical excitation prepares either single eigenstates or coherent superpositions of eigenstates, which may be approximated by zero-order bright states. In the former case, the decay of the excited level is exponential and purely radiative, with the overall fluorescence yield $Q_f = 1$ even when the $s \rightarrow s'$ and $\ell \rightarrow \ell'$ fluorescence components appear in different spectral regions with the yields $Q_f^s + Q_f^\ell = 1$. In the latter case, the decay of the excited state, monitored by the intensity dependence of its fluorescence components, is not purely exponential (modulated by quantum beats or quasi-multiexponential). The overall fluorescence yield is equal to one and the mixing of bright and dark excited states implies an *increase* of excited-state lifetimes with respect to those of pure bright states.

We define a *large molecule* as a system of discrete levels coupled to the continua or quasi-continua, i.e., sets of discrete levels so dense that the level spacing is smaller than the intrinsic level widths. This condition corresponds to the suppression of recurrences, i.e., to irreversibility of transitions. If the continua (quasi-continua) are sets of nonradiant states, the lifetimes of bright states are *shortened* by the coupling to the continuum and the fluorescence yield is reduced, $Q_f < 1$.

If the coupling of a discrete bright state to the continuum is so strong that its coupling to another discrete state may be neglected, its decay rate $s \rightarrow m$ is determined by the Fermi Golden Rule and is proportional to the square of the V_{sm} coupling constant and to the m-level density ρ_m. This situation is called the *large-molecule limit*.

In a large class of molecular systems, the coupling between discrete levels cannot be neglected and the system is described by the *sequential-coupling scheme*: the sequential coupling of the bright state to a limited set of the discrete dark states coupled in turn to the continua or quasi-continua. The time evolution of such systems is a many-step process, the details of which depend on the coupling strength between different groups of states. The decay rates are then limited by that of the slowest step.

1.6 Franck–Condon Rule and Gap Laws

In the previous sections, we focused our attention on energy and time properties of excited molecular states. The time-independent functions of spatial coordinates are also an important parameter in the theory of relaxation processes.

1.6.1 Franck–Condon Rule

The Franck–Condon rule is among the cornerstones of molecular spectroscopy. It was first formulated by Franck [21] in terms of classical physics, based on the assumption that the positions and momenta of the heavy particles (nuclei) in a molecule are conserved during the optical excitations of the electron. The quantum mechanical formulation of the Franck–Condon principle is due to Condon [22] and may be described as follows.

The transition between the initial state $|i\rangle$ and the final state $|f\rangle$ described by the nuclear-electronic wave functions $\Psi_i(q, Q)$ and $\Psi_f(q, Q)$ (q, electronic, and Q, nuclear coordinates, respectively) is induced by a perturbation $V(q, Q)$. In first order, the probability of transition is given by

$$P_{if} = |T_{if}|^2 = |\langle \Psi_i(q, Q)|V(q, Q)|\Psi_f(q, Q)\rangle|^2, \tag{1.63}$$

where the integration is over electronic and nuclear coordinates. If the Born–Oppenheimer factorization is valid in both the initial and the final state, i.e., $\Psi_{i/f} \approx \psi_{i/f}(q, Q)\chi_{v_{i/f}}^{(i/f)}(Q)$, we can write

$$\begin{aligned}
P_{if} &\approx |\langle \psi_i(q, Q)\chi_{v_i}^{(i)}(Q)|V(q, Q)|\psi_f(q, Q)\chi_{v_f}^{(f)}(Q)\rangle|^2 \\
&= |\langle \chi_{v_i}^{(i)}(Q) |V_{if}(Q)| \chi_{v_f}^{(f)}(Q)\rangle|^2, \tag{1.64}
\end{aligned}$$

where (second line) the integration over the electronic coordinates has been carried out and V_{if} is the resulting electronic transition moment. Assuming further that the perturbation V_{if} depends only weakly on Q on the scale of the nuclear vibrations (e.g., on the scale of the zero-point vibrations, which may be small when $|i\rangle$ is the ground state of a rigid molecule) we can write

$$P_{if} \approx |V_{if}(\bar{Q})|^2 |\langle \chi_{v_i}^{(i)}(Q)|\chi_{v_f}^{(f)}(Q)\rangle|^2, \tag{1.65}$$

where $\bar{Q}$ is the average or centroid value corresponding to the range of Q values that contribute mainly to the vibrational overlap integral. The square of the latter, $|\langle \chi_{v_i}^{(i)}(Q)|\chi_{v_f}^{(f)}(Q)\rangle|^2$, is the Franck–Condon factor—the square of the overlap integral of the vibrational wave functions.

This type of approximation is widely used in electronic spectroscopy. For instance, if the transition between the initial and the final state is induced by interaction with the electromagnetic field, the transition moment V_{if} corresponds to the electronic dipole transition moment and is usually denoted μ_{if}, so that the optical transition probability becomes

$$P_{if} \approx |\mathcal{E}\mu_{if}(\bar{Q})|^2 \ |\langle \chi_{v_i}^{(i)}(Q)|\chi_{v_f}^{(f)}(Q)\rangle|^2. \tag{1.66}$$

The intensity distribution in a vibrational progression is thus determined by the dependence of the Franck–Condon factor on the difference $\Delta v = |v_i - v_f|$ of the vibrational quantum numbers in the initial and final states. For small Δv, these values depend on the details of the potential energy surfaces. For example, if the initial and final potential energy surfaces have their minima at different nuclear geometries, an oscillatory behaviour of the Franck–Condon factor will occur. However, for large values of Δv, a more or less smooth decrease of P_{if} is expected, e.g., when the transition takes place between the ground vibrational level, the wave function of which is of Gaussian form centered near $Q \approx 0$, and rapidly oscillating vibrational wave functions associated with a dissociation continuum.

In certain other processes, such as molecular collisions, one component of Q, say Q_k, ranges essentially from zero to infinity. The perturbation $V(q, Q)$ then cannot be approximated by a constant value independent of Q, but on the contrary is found to be strongly dependent on Q because it will differ from zero only in a range where the collision partners are close to each other, for $Q_k \leq Q_{k0}$, where Q_{k0} is the range of the intermolecular interaction.

1.6.2 Energy and Momentum Gap Laws

In a large number of molecular systems, the rate k_{if} of the electronic relaxation may be correlated (all other parameters being identical) with the energy difference ΔE_{if} between the initial state, $|i\rangle$, and the final state, $|f\rangle$. This correlation was noticed initially in an early study of the $T_1 \rightarrow S_0$ intersystem crossing in a series of aromatic polynuclear hydrocarbons [23–25] and is approximated by

$$k_{if} \approx A e^{-\alpha \Delta E_{if}}. \tag{1.67}$$

Similar relationships have been found experimentally in the cases of vibrational relaxation of small molecules in low-temperature matrices [26, 27], of predissociation of molecular complexes [28, 29] and of the vibrational relaxation in gas-phase collisions [30]. An energy gap law is known to operate also in vibrational autoionization. This latter process occurs in excited molecules possessing a weakly bound electron, whose binding energy corresponds to one or a few vibrational quanta of energy [31]. If the molecular ion core is vibrationally excited, it may yield some of its internal vibrational energy to the electron, allowing the latter to escape into the electronic continuum. Vibrational autoionization tends to proceed by transfer of the minimum amount of vibrational energy necessary for ionization to become possible.

Although the term energy gap law is used more widely than momentum gap law, the dependence of the decay rates on the energy gap is not direct but can be traced back to the nodal structures and wavelengths of the wave functions involved in the relaxation process, as implied by Eq. 1.65. In many instances, but not always, the relaxation process depends mainly on the initial and final vibrational wave functions. The gap laws for nonradiative transitions then are related to the Franck–Condon rule. We shall illustrate this by a few characteristic examples. More detailed considerations will be presented in the subsequent chapters.

Intramolecular Electronic Relaxation

Let us consider the rate of relaxation of the vibrationless level $v_i = 0$ of the excited electronic state $|i\rangle$ with energy E_i above the vibrationless ground state. In an isolated molecule, energy and angular momentum must be conserved and therefore the only allowed relaxation channel corresponds to a transition from v_i, J_i to the isoenergetic levels of the electronic ground state, v_f, $J_f = J_i$. The coupling between the electronically excited and the ground state is always a nonadiabatic interaction, which, however, may take various forms. For the present purposes, we shall assume that a single coordinate Q_k corresponding to a nondegenerate vibrational mode dominates the coupling and that the nonadiabatic coupling is vibronic and arises from the nuclear momentum coupling operator.

Vibronic coupling between electronic states n' and n gives coupling terms of the form (in units energy/hc)

$$-2\left(\frac{h}{8\pi^2\mu c}\right) V_{n'n}(Q_k)\left\langle v_{k'}^{(n')}\left|\frac{\partial}{\partial Q_k}\right|v_k^{(n)}\right\rangle, \tag{1.68}$$

where $V_{n'n}(Q_k)$ is the electronic nuclear momentum coupling integral, i.e., $V_{n'n}(Q_k) = \langle\psi_{n'}|\partial/\partial Q_k|\psi_n\rangle$. Note that the Q_k's here are not assumed to represent mass-weighted normal coordinates but instead have the dimension of length. μ is the appropriate reduced mass. If the dimensions of mass are in amu and displacements in a_0, then the factor $h/8\pi^2 c$ amounts to 60.1997 amu a_0^2 cm^{-1}. The nuclear momentum coupling function $V_{n'n}(Q_k)$ cannot be assumed to be constant or depend only weakly on Q_k, and Eq. 1.65 can therefore not be used. We now show that nuclear momentum coupling nevertheless obeys an energy gap law.

The vibrational factor of Eq. 1.68 can be evaluated explicitly if the potential surfaces in the excited state and in the ground state are the same and are harmonic. It is then given by

$$-2\frac{h}{8\pi^2\mu c}\left\langle v_{k'}^{(n')}\left|\frac{\partial}{\partial Q_k}\right|v_k^{(n)}\right\rangle = \sqrt{\frac{h}{8\pi^2\mu c}\omega_k}\sqrt{v_k \pm 1}, \tag{1.69}$$

where ω_k is the harmonic frequency of mode k in cm^{-1}. Equation 1.69 indicates that the electronic relaxation, when driven by nonadiabatic nuclear momentum coupling, will proceed with the minimum nonzero amount of exchange of energy between the electronic and nuclear degrees of freedom, namely $|\Delta v| = 1$. When the potential surfaces of the interacting electronic states are not harmonic and in addition are different, the selection rule of Eq. 1.69 will obviously be relaxed, but the preference for minimum exchange of energy between the light and heavy particles will be preserved. Although it is not obvious that the energy gap law will take exactly the form of Eq. 1.67, the latter expression certainly provides a reasonable parametrization.

Predissociation

Consider the $v_k = 0$ level of an excited electronic or vibrational state of a molecule XY*, which is predissociated by its coupling to the continuum of free (i.e., lying above the dissociation threshold) X + Y states associated with the ground state of XY. If the predissociation is electronic, the same arguments apply as in the preceding subsection and point to the existence of an energy gap law. If the predissociation is vibrational, the coupling arises from anharmonic potential terms proportional to $Q_i^n Q_f^m$ ($m + n \geq 3$). The vibrational factor in Eq. 1.68 must then be replaced by the appropriate higher order potential terms. These in turn can be constructed by matrix multiplication from dipole integrals $\langle v_k' | Q_k | v_k \rangle$, which, in the harmonic approximation, are known to be

$$\langle v_k' | Q_k | v_k \rangle = \sqrt{\frac{h}{8\pi^2 \mu c \omega_k}} \sqrt{v_k \pm 1}. \tag{1.70}$$

The analogy with Eq. 1.69 is evident and shows that an energy gap law is operational also in vibrational (Fermi-type) coupling.

Figure 1.8 (left) illustrates another aspect of the energy gap law. Inspection of the wave functions shown there indicates that the rate of predissociation is limited by the small value of the overlap integral between the dramatically different vibrational wave functions $v_k = 0$, bound and localized near the equilibrium geometry, and $E = E_k$, free and resembling a free-particle wave. Obviously, whatever the precise coupling mechanism is, the transition integral originates in the limited region where the $v_k = 0$ wave function differs appreciably from zero, and it is conditioned by the difference of the effective wavelengths of the two functions. As the kinetic energy release of the predissociation process increases, the wavelength difference becomes larger and the transition integral will become smaller due to the rapid oscillations of the product of the two functions (Fig. 1.8 (right)). We discuss this aspect in more detail in the following subsection.

Collisional Relaxation

Consider an inelastic collision between two reactants A and C. During this process, energy is transferred from (or to) the internal degrees of freedom of A and/or C to (or from) the translational energy of the pair. We limit our considerations once again to a single nondegenerate coordinate, the reaction coordinate Q. According to Eq. 1.64, we must evaluate transition integrals of the form

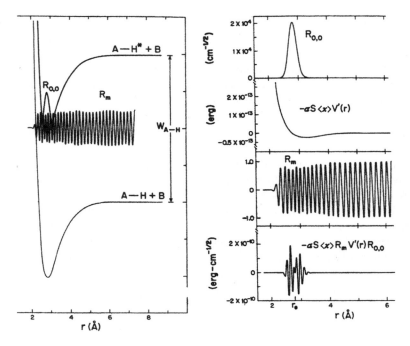

Fig. 1.8. At left: the schematic representation of the potential energy curves and wave functions of the XY* and X + Y states; at right: the r-dependence of the $v = 0$ function R_{00}, of the coupling strength $(\alpha S(x)V(r))$ of the X + Y function (R_m) and of their triple product (from Ref. [29])

$$T_{if} = \int \chi_f^{(f)}(Q)\, V_{if}(Q)\, \chi_i^{(i)}(Q)\, dQ. \tag{1.71}$$

The continuum vibrational wave functions are conveniently written in phase-amplitude form,

$$\chi(Q) = \sqrt{\frac{1}{\pi}}\sqrt{\frac{2\mu}{\hbar^2}}\, \alpha(Q)\sin[\varphi(Q)], \tag{1.72}$$

which explicitly brings out their oscillatory structure. The phase function $\varphi(Q)$ is related to the amplitude function $\alpha(Q)$ according to the relation

$$\varphi(Q) = \int_0^Q \frac{1}{\alpha^2(Q')}\, dQ',$$

which quantifies the intuitively familiar fact that the phase accumulation as function of Q is the faster the smaller the amplitude [32]. The amplitude function in turn can be evaluated in the WKB approximation as $\alpha(Q) = \left\{ [2\mu/\hbar^2]\,[E - U(Q)] \right\}^{-1/4}$, where $U(Q)$ is the potential energy. These continuum functions have dimension [energy]$^{-1/2}$[length]$^{-1/2}$ and they are energy normalized in the sense that

$$\int \chi_f^{(f)} \chi_i^{(i)} dQ = \delta(E_f - E_i).$$

Asymptotically the potential $U(Q)$ approaches a constant value and one has $\varphi(Q) \to kQ + \Delta(E)$, corresponding to a translational energy $E = \hbar^2 k^2 / 2\mu$ and a constant amplitude

$$\alpha = \sqrt{\frac{1}{k}}. \tag{1.73}$$

$\Delta(E)$ is the elastic scattering phase shift due to the potential $U(Q)$. Note that the amplitude factor Eq. 1.73 decreases according to $E^{-1/4}$ and thus acts in accordance with the energy gap law.

The transition integral Eq. 1.71 now becomes

$$T_{if} = \frac{2\mu}{\pi \hbar^2} \int \alpha_f(Q) \sin[\varphi_f(Q)] V_{if}(Q) \alpha_i(Q) \sin[\varphi_i(Q)] \, dQ. \tag{1.74}$$

Dropping the amplitude/normalization factor, we can further rewrite [33] the product of the two vibrational functions in Eq. 1.74 as

$$\chi_f^{(f)}(Q)\chi_i^{(i)}(Q) = \frac{1}{2} \left\{ \cos[\varphi_f(Q) - \varphi_i(Q)] - \cos[\varphi_f(Q) + \varphi_i(Q)] \right\}. \tag{1.75}$$

The second term on the right-hand side of this formula generally oscillates rapidly and will contribute little to the integral Eq. 1.71. Therefore, the integral is nonnegligible if there is a range of Q values where

$$V_{if}(Q) \cos[(\varphi_f(Q) - \varphi_i(Q)] \tag{1.76}$$

varies smoothly with Q rather than oscillating. If $V_{if}(Q)$ itself is smooth, the condition for this to be the case is that $\varphi_i \approx \varphi_f$ over a range, i.e., the relative phase of the two waves must be *stationary*. If, on the other hand, the interaction V_{if} is restricted to a small region near $Q \approx 0$ (reaction region), the transition integral will be nonvanishing only when the phase of the integrand is stationary in that same region. The outcome in all events is that the linear moments and energies of the coupled states cannot be too different for efficient relaxation to occur, i.e., the *momentum (energy) gap law* will be effective. Phase-amplitude functions can also be defined in the bound-state range (but we shall not discuss this in detail here), and therefore the interpretation of the energy gap law in terms of a stationary phase condition is valid in general.

2

Experimental Techniques

2.1 Principles of Time- and Energy-Resolved Experiments

The scope of this chapter is a rapid survey of experimental methods used for the studies of the time-evolution and relaxation of electronically and/or vibrationally excited molecules. The success of these studies is due to the progress in the molecular spectroscopy conditioned by the development of the laser light sources, of the ultrarapid electronics and computer technology. We will focus our attention on the methods of the laser spectroscopy.

As previously discussed, the dynamics of an excited molecule depends on the level pattern of the initially prepared state. The information about the dynamics may be thus deduced either from the *energy-resolved* absorption spectrum or from a direct *time-resolved* observation of the time evolution when the excited state is prepared by a short-pulse excitation. So, the homogeneous width γ_i^{hom} of the level $|i\rangle$ in the absence of dephasing effects is related to the decay rate of its population k_i as

$$\gamma_i^{\text{hom}} = \hbar k_i, \qquad (2.1)$$

so that the widths of ~ 1.6 MHz ($\sim 5 \times 10^{-5}$ cm^{-1}) and ~ 160 MHz ($\sim 5 \times 10^{-3}$ cm^{-1}) correspond, respectively, to $k_i = 10^7$ and 10^9 s^{-1}. Also, the frequency of quantum beats between two coherently excited i- and j-levels ω_{ij} is determined by the energy gap, ΔE_{ij}.

The photophysical and photochemical processes taking place in the real-life conditions: room-temperature solutions and gaseous mixtures at the normal (~ 1 bar) pressure are of great interest. Unfortunately, in these conditions, the excited molecular states are strongly perturbed by interactions with their environment; their decay is accelerated and their spectra are broadened by population and phase relaxation due to the solvent movements. In order to separate intramolecular and intermolecular processes, it is necessary to start by the study of *isolated molecules* and then investigate the effects of *molecular interactions*.

An excited molecule may be considered as isolated when no interaction with its environment occurs during a time long as compared to its decay time. At $T \approx 300$ K, collision rate in the vapor composed of molecules with molecular masses of $M \approx 40$ is of the order of $k_{coll} = 10^{10} \text{ s}^{-1}\text{bar}^{-1}$, so that the excited molecule with decay rate $k_i \approx 10^7 \text{ s}^{-1}$ is collision-free at the gas pressure $p \leq 10^{-4}$ bar ($k_{coll} \times p = 10^6$ s^{-1}), whereas for a molecule with $k_i \approx 10^9 \text{ s}^{-1}$, the pressure of $p \leq 10^{-2}$ bar is low enough

The environment effects (the collisional relaxation and collisional broadening of energy levels) are eliminated in the low-pressure gas at room temperature but the spectral resolution is still limited. The spectra of medium-sized and large molecules cannot be completely resolved in view of the spectral congestion due to overlapping transitions from a large number of thermally populated rotational and vibrational levels. The spectral congestion is reduced by the vibrational and rotational cooling in *supersonic expansions*: the spectra are thus dramatically simplified by the suppression of hot vibronic bands and narrowing of the rotational envelopes of cold bands.

The principal source of the inhomogeneous broadening of spectral lines is the Doppler effect, the Doppler width being proportional to the frequency of transition, ω,

$$\Delta\omega^{\text{Dop}} = \frac{v_z}{c}\omega, \tag{2.2}$$

where v_z is the projection of the velocity vector $\mathbf{v}$ on the light beam axis and c is the velocity of light. For $M = 40$ and $T = 300$ K, the average velocity, $\langle v \rangle \approx 300$ m/s, equation (2.2) implies $\Delta\omega^{\text{Dop}} \approx 3 \times 10^{-2}$ and 10^{-3} cm^{-1} in the 30,000 and 1,000 cm^{-1} spectral ranges, respectively. On the other hand, the radiative decay rates and radiative widths of excited levels vary (in view of the ω^3 factor in the Einstein coefficient of spontaneous emission) from (typically) $5 \times 10^{-4} \text{ cm}^{-1}$ for $\omega \approx 30,000$ cm^{-1} to 10^{-7} cm^{-1} for 1,000 cm^{-1}. The Doppler widths exceed the intrinsic line widths so that the fine-structure measurements necessitate elimination or reduction of the Doppler broadening by one of the techniques of the sub-Doppler spectroscopy (see Sect. 2.2.2)

In order to understand the medium effects in dense phases, it is convenient to investigate model systems in which these effects are weak and may be varied in a controlled way. The gas-phase collisions $A^* + C$ with a variable concentration of different collision partners C, half-collisions (dissociation of A^*C complexes), spectroscopy of molecules trapped in well-defined sites of rare-gas crystals (matrices) and clusters will be briefly discussed in Sect. 2.4

The best tools for the energy, as well as for the time-resolved studies, are the coherent light sources (such as single-mode and mode-locked lasers) with a high simultaneous time and energy resolution limited only by the Fourier relation,

$$\delta\omega_{\text{rad}}[\text{cm}^{-1}] \times \delta t_{\text{rad}}[\text{s}] \approx 5.3 \times 10^{-12}\text{cm}^{-1} \times \text{s}. \tag{2.3}$$

So, in order to attain energy resolution of 10^{-5} cm^{-1}, the laser intensity must be maintained constant during at least 10^{-6} s, whereas the spectral width of a 100-fs pulse cannot be reduced below $\delta\omega_{\text{rad}} \geq 50 \text{ cm}^{-1}$.

In view of these limitations, the experimental methods may be roughly divided into the studies of the fine structure of the absorption spectra with the high *energy resolution* and no (or limited) time resolution and the *time-resolved studies* with a limited energy resolution. Both types of measurements are feasible in a wide frequency range, but in view of the dependence of k_i and $\Delta\omega_{Dop}$ on ω, the high-resolution spectroscopic studies are easier in the infrared, whereas, in view of the short lifetimes of electronically excited species, the time-resolved experiments in collision-free conditions are easier in the ultraviolet/visible spectral range.

2.2 Isolated Molecules

2.2.1 Gases and Supersonic Jets

Gas Cells

The collision-free conditions for molecules with a decay rate of 10^7 s^{-1} are attained in the room-temperature gas $p \leq 10^{-4}$ bar, which corresponds to the concentration c_A of $\sim 5 \times 10^{-6}$ mole/l and a very low optical density. The concentrations of the species with low vapor pressure are still lower unless the temperature is increased, which implies an enhanced spectral congestion. A large class of thermally unstable or reactive species are excluded from gas-cell experiments. The interest of the gas-cell experiments is a possibility of precise measurements of the concentrations (partial pressures) of the species under study (A) and of its collision partner (C).

Supersonic Jets

The supersonic jets developed since the early 1970s mark a major progress in the study of medium-sized molecules (cf. [1]). When the rare gas C (He or Ar) containing a small fraction of the compound A is expanded through a small hole into a strongly pumped vacuum chamber, the collisions between molecules and carrier-gas atoms taking place in the initial part of the supersonic expansion transform the major part of their internal energy into that of the linear movement along the jet axis. The average velocity in its final part is nearly the same for A and C in spite of their different masses so that number of collisions is strongly reduced, which allows attaining collisionless conditions at a density of A higher than in gas cells. The distribution width of final velocities is a measure of the translational temperature, T_{tr}, of the jet, which is typically of the order of $1 K$.

The internal energy of molecules, roughly described by the vibrational (T_{vib}) and rotational (T_{rot}) temperatures, tends to the equilibrium with T_{tr}, but in view of a limited number of collisions, this equilibrium is not attained. The rotational temperature, T_{rot}, deduced from the population ratios of low J-levels is slightly higher than T_{tr}, whereas T_{vib} is of the order of 30 K for medium-sized molecules. The absorption spectrum is simplified by depopulation of higher vibrational and rotational levels, i.e., suppression of hot bands and narrowing of rotational envelopes of "cold" vibronic bands (Fig. 2.1).

The compounds with very low vapor pressures or thermally labile may be injected into the expanding gas using such techniques as the laser evaporation [2]. The strongly reactive radicals or ions introduced or created within a jet survive in the absence of collisions during the flight time of several microseconds.

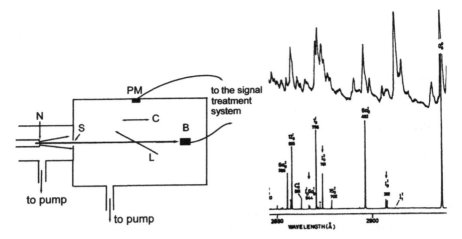

Fig. 2.1. (Left) A scheme of a skimmered jet device: N, nozzle; S, skimmer; L, a laser beam crossed with the jet; C, condensing lens; PM, photomultiplier fluorescence detector; B, bolometer thermal detector. (Right) The fluorescence-excitation spectrum of aniline in the room-temperature gas (top) and in the supersonic jet (bottom) (from [26])

The free supersonic jet may be skimmered: its central part, composed of particles with velocity vectors parallel to the jet axis, penetrates through a small aperture (skimmer) from the expansion chamber to a second, high-vacuum chamber and forms a nearly parallel molecular beam.

2.2.2 The Spectroscopy of Isolated Molecules

High Sensitivity Techniques

The main difficulty encountered in the spectroscopy of low-pressure gases and supersonic jets is the low concentration of molecules $A - c_A$ and the small length of the optical path d, limited by the dimensions of the cell (a few cm) and of the jet (a few mm). The accuracy of measurements of the optical density $OD(\omega) = \log[I_0(\omega)/I(\omega)] = c_A \varepsilon_A(\omega)d$ (where I_0 and I are intensities of the incident and transmitted light and $\varepsilon(\omega)$ is the absorption coefficient) depends on the relation between the $\Delta I = I_0(\omega) - I(\omega)$ absorption signal proportional to I_0 and the photon noise, which is proportional to $\sqrt{I_0}$. The signal-to-noise ratio decreases when the light intensity is reduced in order to avoid saturation effects and the OD measurements are imprecise even for relatively large ε values. The $I_0(\omega)/I(\omega)$ ratio may be improved by increasing the optical path in multireflection cells [3] or by using such techniques as the cavity ring-down spectroscopy [4]. Nevertheless, it is often necessary to replace the direct absorption measurements by the techniques of the *action spectroscopy* in which the absorption spectrum is deduced from the variation of the excited-state population with the excitation frequency ω_{exc} [3].

For a large class of organic molecules (aromatics, hetero-aromatics) with high fluorescence yields in the ultraviolet/visible spectral range, the *fluorescence-excitation spectroscopy* is widely used. The intensity of fluorescence emitted at a constant frequency ω_{obs} at a constant delay with respect to the laser pulse is recorded while ω_{exc} is varied. In optically thin ($OD \ll 1$) samples and in the absence of saturation, the intensity of fluorescence emitted upon the laser irradiation is proportional to the optical density, fluorescence yield and laser intensity at ω_{exc}. If the frequency dependence of the laser intensity $I_0(\omega)$ and of the fluorescence yield $Q(\omega)$ is known, the shape of the absorption spectrum $OD(\omega)/OD(\omega_0)$ may be calculated but absolute values of the optical density cannot be determined.

If the fluorescence yield of the A^* state is too low, the action spectra may be recorded by an *active probing* of its population (the *pump-probe* techniques). When a higher state A^{**} has a larger emission yield or dissociates with formation of strongly fluorescent fragments, the population of the A^* state is deduced from the intensity of the emission due to the $A^* \rightarrow A^{**}$ excitation by a second (probe) laser pulse (*laser-induced-fluorescence, LIF*) (Fig. 2.2 (left)) [5]. Another pump-probe technique consists in the determination of the A^*-level population by REMPI (resonance-enhanced multiphoton ionization): the excited molecules XY^* are ionized by the one-, two- or many-photon absorption,

$$XY^* + n\hbar\omega \rightarrow XY^+ + e^- \quad \text{or} \quad X^+ + Y + e^-.$$

This method is sensitive because of a high efficiency of electron and ion collection and detection and allows identifying the presence of dissociation products by a separate counting of XY^+ and X^+ ions differing by their masses (Fig. 2.2 (right)) [6].

The fluorescence excitation spectroscopy is not well adapted to the vibrational spectroscopy in infrared because of a strong collisional quenching of the long-lived vibrational fluorescence. The LIF or REMPI determination of vibrational-level pop-

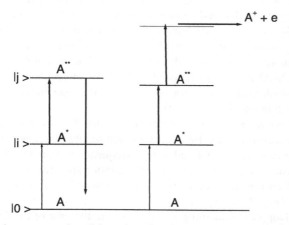

Fig. 2.2. Schematic representation of detection of excited species by laser-induced fluorescence (left) and multiphoton ionization (right)

ulation using the UV probe after infrared excitation is possible but is still not very widely used [7, 8].

Besides the Fourier-transform (TF) infrared spectroscopy, the high-resolution spectra of nonfluorescent compounds have been obtained using the skimmered molecular beam crossed with a light beam issued from a tunable laser scanning the absorption range. The signal of a thermal detector (bolometer) placed at the beam axis in the high-vacuum chamber is recorded as a function of the laser frequency ω_{las}. The bolometer signal is enhanced when the energy content of a molecule is increased by absorption in the case of resonance $\omega_{las} = \omega_{abs}$. On the other hand, upon the excitation of dissociative or predissociated level, the dissociation products are rejected from the beam and the detector signal is reduced [1].

Sub-Doppler Spectroscopy

Among a large number of techniques developed in order to get rid of the Doppler broadening, we will give only a few examples.

The Doppler widths of absorption lines are strongly reduced when weakly divergent laser and molecular beams cross at a 90° angle. The projection of the molecular velocity vector on the laser beam axis z is then close to zero: $v_z = 0$. The Doppler width is reduced to 1/30 of that measured in the gas for the beam divergence of 2° (cf. [9]).

One of the most important techniques of the sub-Doppler spectroscopy is the *optical hole burning* [10, 11]. An intense narrow-band pump laser with a frequency ω_{las} fixed inside the inhomogeneous line profile excites a limited class of molecules with the v_z value such that $\omega_{abs} + (v_z/c)\omega_{abs} = \omega_{las}$. The ground-state population of this class of molecules is reduced and the sample becomes transparent in the energy band centered at ω_{las} with a width equal to the homogeneous width of the absorption line (Fig. 2.3 (a)). The shape of this hole is measured by scanning a weak, narrow-band probe laser across the band. The hole is filled by relaxation of excited molecules and by the collisional reorientation of velocity vectors of the ensemble of molecules. Its duration is thus determined by the lifetime of the excited level and by the pressure-dependent collision rate [10, 11]. This technique may also be applied to rigid solutions (mixed crystals, glasses) at $T \approx 4$ K. The holes created in these media are long lived in view of reduced mobility of molecules or even permanent in the case of the photochemical processes such as the laser-induced dissociation or isomerization of a molecular species [12].

The simplified version of the hole burning is the *fluorescence line narrowing, FLN*: the fluorescence excited by a narrow-band tunable laser is detected across a narrow-slit monochromator centered at ω_{det} frequency within an inhomogeneously broadened emission band. The fluorescence-excitation spectrum recorded in this way is composed of narrow lines corresponding to absorption of a class of molecules emitting at ω_{det} (Fig. 2.3 (b)) [13].

At last, the Doppler broadening is suppressed in the case of the fluorescence excitation by the simultaneous absorption of two photons issued from two anticollinear laser beams with the same frequency, $\omega_{las} = \omega_{abs}/2$. For a molecule with projection

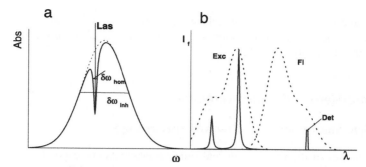

Fig. 2.3. (Left) Schematic representation of a hole burnt in an inhomogeneously broadened absorption line. (Right) Scheme of the fluorescence excitation spectra observed with broadband detection (broken line) and the excitation spectrum observed with a narrow-band detection, Det (solid line)

of the velocity vector on the laser axis v_z, the frequency of the first laser beam is Doppler shifted to $\omega_{las} + v_z \omega_{las}/c$ and that of the second laser to $\omega_{las} + v_z \omega_{las}/c$, so that these shifts compensate [14].

External Field Effect

In the case of a weak coupling between two ro-vibronic levels, $|s, J\rangle$ and $|\ell, J\rangle$, belonging to different electronic states (S_1 and T_1 or S_1 and S_0) such that the coupling constant is small compared with the energy gap, $V_{s\ell} \ll |E_s - E_\ell|$, the detection of spectral perturbations is difficult. In order to measure the $V_{s\ell}$ and $|E_s - E_\ell|$ parameters, it is convenient to bring a pair of levels to the resonance in an external magnetic (or electric) field. The resonance is evidenced by the change in the fluorescence lifetime and yield resulting from the strong s-ℓ mixing in the field corresponding to the level crossing B_c (or E_c) (Fig. 2.4). If the magnetic (or electric) dipole moments of the $|s\rangle$ and $|\ell\rangle$ states are known, the energy of the dark $|\ell\rangle$ level in the zero field may

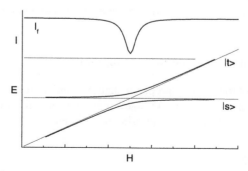

Fig. 2.4. Crossing of $J_t = J_s$ levels in the magnetic field and the corresponding line in the level-anticrossing spectrum

be calculated whereas $V_{s\ell}$ may be deduced from the width of the anticrossing signal. Note that all these data are Doppler free, the interaction taking place between the levels of the same molecule [15].

2.3 Time-Resolved Measurements

2.3.1 Light Sources for Time-Resolved Spectroscopy

The milestones of the progress in real-time experiments correspond to the appearance of light sources with time resolution corresponding to different time scales of molecular processes:

- The flash lamps with ~ 1 ms pulses built in the 1950–60 period allowed the study of strongly forbidden transitions, such as the $T_1 \rightarrow S_0$ relaxation with decay times in the 10^{-3}–10^1 s^{-1} range,
- Around 1962, there appears the first generation of solid-state (ruby, Nd-glass) and gas (N_2^+, CO_2, Ar, Kr, ...) lasers with the 5–30-ns light pulses at several fixed frequencies whereas the light detectors with the nanosecond time resolution are imported from the nuclear to the molecular physics. The studies of the excited singlet states S_1 in solutions are rapidly developed.
- The real revolution in the studies of isolated molecules started in 1967 with appearance of tunable dye lasers, which made possible the selective excitation of individual energy levels. Other tunable light sources based on the electron (positron) accelerators (synchrotron radiation, free-electron lasers) have been developed since 1975.
- The time resolution necessary for the study of ultrafast processes, such as the intramolecular electron or proton transfer and isomerization with subpicosecond rates, was attained by a recent development of solid-state (e.g., Ti/Al$_2$O$_3$) lasers with a pulse duration of a few femtoseconds.

It is thus possible to prepare excited states A* in a wide energy range with a laser-pulse duration that fulfills the condition $\delta t \ll \tau_{A^*}$ (where δt is the light-pulse duration and τ is the characteristic time of an excited-state evolution) for almost all molecular systems.

2.3.2 Monitoring of the Excited-State Evolution

The time evolution of the initially prepared state A* (and, if necessary, of other levels populated by the relaxation of the A* state) may be monitored by the methods that we divide into active and passive monitoring techniques.

Fluorescence Decay

The simplest passive method is the measurement of the fluorescence intensity, $I_f(t)$, as a function of the delay time, $\Delta t = t - t_{\text{pump}}$, assuming that I_f is proportional to

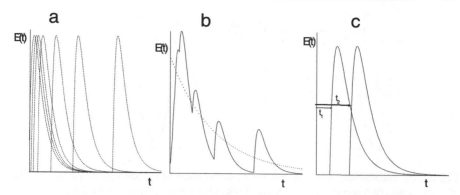

Fig. 2.5. (a) The photon distribution in the fluorescence decay, (b) the integrated signal and its fit by a single-exponential decay, (c) principle of the photon counting

the population of the A* level. This assumption is valid in the absence of a strong stimulated emission and of the radiation trapping.

The time resolution of the fluorescence detector, δt_{det}, is limited by the time width of the electric signal induced by detection of a single photon. This width is of the order of a few nanoseconds in the case of rapid photomultipliers. When the fluorescence decay time, τ, is long compared with δt_{det} and the average number of photons emitted after each laser shot, $\langle n_{ph} \rangle$, is so large that the electric pulses due to different single-photon events overlap, the shape of the electric signal $E(t - t_{pump})$ may be considered as equivalent of the time dependence of the fluorescence intensity $I_f(t - t_{pump})$. The decay rate is then estimated by fitting the observed curve $E(t - t_{pump})$ averaged for a number of laser shots by a monoexponential or multiexponential function (Fig. 2.5 (b)). The accuracy of the measurement is limited by the signal-to-noise ratio decreasing with the decrease of $\langle n_{ph} \rangle$ values. For $\langle n_{ph} \rangle \leq 1$, it is thus preferable to switch to the *correlated single-photon counting* technique. The fluorescence intensity is maintained at the $\langle n_{ph} \rangle \ll 1$ level so that no more than one photon is recorded after each laser shot. The delay between t_{pump} and the rising edge of the detector one-photon signal, $t_{ph} \, \Delta t = t_{ph} - t_{pump}$, is measured for each photon (Fig. 2.5 (c)) and the decay curve builds up as the time distribution of delays: $n_{ph}(\Delta t)$. The time resolution (not limited by the duration of the detector signal) attains about 100 ps; this is actually the limit of the passive fluorescence monitoring (cf. however, the fluorescence up-conversion technique described at the end of the next section).

Pump-Probe Techniques

A higher time resolution is attained in the active *pump-probe* technique in which the excited system prepared by a strong *pump pulse* is interrogated by a weaker *probe pulse* applied with a variable delay. The time resolution of these techniques is not limited by the detector but only by duration of the pump and probe pulses and by the accuracy of the measurement of the $\Delta t = t_{probe} - t_{pump}$ delay.

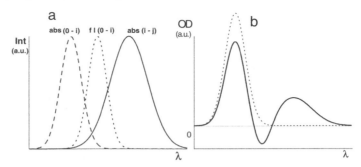

Fig. 2.6. (a) The absorption 0-i, fluorescence i-0 and transient i-j absorption spectra. (b) The effective absorption spectrum at $t < t_{\text{pump}}$ (broken line) and $t \geq t_{\text{pump}}$ (solid line)

When the $A \rightarrow A^*$ absorption is strong and a nonnegligible fraction of molecules is excited by the pump pulse, the time dependence of their populations may be checked by measurements of the transmission $T(\omega, t - t_{\text{pump}})$ of the probe pulse in different spectral regions. The transmission is reduced ($\Delta T < 0$) in the spectral range of the transient absorption of the A^* molecules and increased in those of the ground-state absorption (bleached by the ground-state depopulation) and of the $A^* \rightarrow A$ emission (in which the light signal is enhanced by the stimulated emission) (Fig. 2.6). The decay rate of the excited state population is then given by the time dependence of the $\Delta T(\omega)$ differential spectrum. This technique is currently used for condensed phases and medium-pressure gaseous samples [16].

The concentration of collision-free molecules in a low-pressure gas or in a supersonic jet is too low to give rise to measurable absorption changes. The more sensitive techniques of detection of the excited-state populations are the previously described LIF and REMPI methods: the fluorescence from higher A^{**} levels or the photoionization is induced by a short (ultrashort) probe pulse applied with a variable delay, Δt, and the time-integrated photon or ion signal, $I_\infty(\Delta t)$, is plotted against Δt. Δt may be precisely measured for subpicosecond pump and probe pulses as the difference of optical path lengths Δl in an optical delay line, i.e., $\Delta t = \Delta l / c = 30$ fs/μm.

The up-conversion of the fluorescence signal is a modification of the pump-probe technique. The fluorescence with the frequency ω_{f} is mixed in a nonlinear crystal with an ultrashort pulse of a probe laser ω_{las} delayed by a variable delay, Δt. The intensity I_{sig} of an up-converted signal with the frequency ω_{up} corresponding to the sum of frequencies, $\omega_{\text{f}} + \omega_{\text{las}}$, is then measured. I_{sig} is proportional to $I_{\text{las}} \times I_{\text{f}}(\Delta t)$, yielding the time dependence of the fluorescence intensity $I_{\text{f}}(t)$ [17].

2.4 Model Systems for Molecular Interaction Studies

In dense media, the dynamics of the molecule A is modified by the many-body interaction with other molecules of the same (A) or of the different (C) species. The strength of this interaction depends on the *intermolecular potential* A–A or A–

C. In view of the complexity of these interactions, the studies of model systems in which the molecular interactions are dominated by two-body events are useful. For the interpretation of experimental data, the two-body A–C potentials calculated by the *ab initio* or semiempirical methods may be applied.

2.4.1 Gas-Phase Collisions

In a typical gas-phase experiment, the molecule A excited to a well-defined vibronic (or ro-vibronic) level A^* collides with a rare-gas atom or a molecule C. The partial pressures of A and C (p_A and p_C) are usually chosen in such a way that the probability of the $A^* + A$ collision during the lifetime of the excited state is close to zero whereas the average number of $A^* + C$ collisions does not exceed one per lifetime. The effects of collisions on the relaxation of the A^* state are determined by measurements of the dependence of the decay rate of the A^* state population on the partial pressure of C for different collision partners, C, and different collision energies.

The information deduced from the studies of $A + C$ collisions is limited by a wide distribution of collision energies, impact parameters and mutual orientations of colliding particles. Only the average values of the A–C potential are accessible. The distribution width may be reduced in crossed-beam experiments [16]. When the beam of excited jet-cooled molecules A^* crosses at the angle θ, a beam of collision partners C, the translational energy in the center-of-mass system is fixed and determined by the beam velocities and the angle θ. The molecules A^* may also be partially oriented prior to the collision by an electric field or by excitation using the polarized light so that the angular distribution of molecular axes is not isotropic. The information about the $A^* + C$ collision is contained in the velocity, angular distribution and ro-vibronic-level populations of $\tilde{A}$ molecules ejected from the beam by the collision. This is, however, a heavy experiment and the interpretation of its results is not simple [18].

2.4.2 Rare-Gas Matrices and Clusters

The rare gases and N_2 condensed at $T \approx 4$ K form cubic crystals in which the guest A occupies well-defined substitution sites AC_n in the infinite host lattice C_∞. The nearly identical structure of each type of sites is evidenced by their quasi-linear absorption (fluorescence-excitation) spectra with line widths of the order of 1 cm^{-1}. In view of the weakness of the guest–host interaction, the level energies and their relaxation times are not very different from those of the free molecule in the gas phase. The information about this interaction may be deduced from the host and site dependence of the level energies, transition moments and relaxation rates upon the narrow-band excitation [19].

The small AC_n ($n = 2$ to ~ 30) clusters formed in supersonic expansions have the properties similar to those of the AC_∞ crystals but their spectra are more complex and/or broadened because of a wide distribution of configurations of finite C_n structures. The free rotation of a guest molecule, with the exception of several hydrides, is blocked in the crystal lattice of rare-gas matrices and clusters, as shown by the absence of the rotational structures in their absorption spectra.

The guest–host interaction is still weaker for molecules contained in the micro-droplets of the superfluid ^{4}He at $T < 1$ K [20, 21]. The droplets obtained by expansion to the vacuum of the gas cooled to ~ 20 K contain typically a few thousand He atoms and are doped with molecules A by passing them through a cell containing the low-pressure vapor. The spectra show a well-resolved rotational structure and only very weak phonon bands accompanying narrow zero-phonon lines with the homogeneous widths corresponding to 0.01–1-ns lifetimes of excited vibrational levels. The apparent change of the free-molecule rotational constants B indicates that, even in the helium superfluid solvent, the rotation is perturbed.

2.4.3 AC Complexes and Half-Collisions

The weakly bound AC complexes with bounding energies D_0 of several to a few thousand cm^{-1} are formed in low-energy A + C + C three-body collisions in super-sonic expansions and cooled to $T_{vib} \approx 5 - 30$ K. The spectroscopy and dynamics of complexes contain important information about the A–C interaction [18].

The electronic and vibrational spectra of the AC complex show usually a well-resolved vibrational structure corresponding to external (intermolecular) vibrational modes. The geometry of the complex, its bonding energy and the rigidity of its structure deduced from its spectrum allow checking the validity of the calculated A–C potential in a limited part of space in the vicinity of the equilibrium configuration.

On the other hand, the dissociation of the AC complex excited to the energy level with $E_{vib} > D_0$: the half-collision $A^*C \rightarrow A + C$ is an analogue of the $A^* + C \rightarrow A + C$ collision. In contrast with full collisions, the initial energy and configuration of the A+C pair is fixed. The dissociation rate as well as the velocity and the vibrational and rotational energy distribution in A+C fragments allow determining the dissociation mechanism. The data obtained for a well-defined structure and energy of A^*C are useful in the analysis of the data from the full-collision experiments with a wide distribution of energies and mutual orientations of A^* and C colliders.

2.4.4 From the Model to Real Systems

The studies of model systems are nothing but appetizers with respect to the problems encountered in the studies of dense phases. The time evolution of an excited system composed of a molecule and its solvation shell is a result of several intramolecular and intermolecular processes. Because the time scales of different processes are different, they may sometimes be separated in experiments with a high time resolution. For instance, the intramolecular electron or proton transfer processes with ~ 50 fs characteristic times seem not to be perturbed by the reorganization of the solvation shell taking place at the 1-ps time scale (cf. [22]).

One can presume that the development of the *single-molecule* spectroscopy will allow studying the structure and time evolution of a well-defined site in the rigid or fluid solvent instead of the analysis of statistical ensembles. Upon a sharp focalization of the laser beam in the volume of the order of λ^3 of a dilute solution, a single molecule contained in this volume may be excited by absorption of one or two photons and followed by a sensitive fluorescence detector [23]. Combined with the near-field microscopy, this technique makes possible determination of the spectrum and dynamics of molecules occupying different sites on the solid surface such as the active sites of heterogeneous catalysis. On the other hand, the microscopic study of the diffusion in the fluid solutions is now possible [24].

The frequency-sum-generation (SFG) techniques allow determining selectively the spectra of molecules submitted to strong field gradients on the solid surfaces in spite of the presence of identical molecules in solution. For instance, the spectra of molecules (ions) of an electrolyte contained in a monomolecular layer on the surface of an electrode are studied in this way [25].

3

Intramolecular Vibrational Redistribution

3.1 Introduction

3.1.1 Vibrational Relaxation and Redistribution

When the molecular system is in equilibrium with its heat bath characterized by a temperature T, the populations N_n of its vibrational levels $|n\rangle$ are given by the Boltzmann law,

$$N_n/N_0 = (g_n/g_0)e^{-(E_n - E_0)/kT}, \qquad (3.1)$$

where E_0 and E_n are the energies of the levels $|0\rangle$ and $|n\rangle$ and g_n and g_0 their degeneracies. A molecule initially prepared with a thermally nonequilibrated population distribution will evolve toward such an equilibrium. This process, involving population redistribution between vibrational levels and energy transfer between the molecular system and its thermal bath—the *vibrational relaxation*—will be discussed in the next chapter. We are interested here in the evolution taking place in an isolated molecule: the *intramolecular vibrational redistribution* (IVR). Because the energy transfer to or from the heat bath is precluded, the energy of the molecule must be conserved and populations may be redistributed only between resonant levels. No IVR will thus occur in a small molecule with distant vibrational levels and a complete randomization of the energy distribution takes place only in the molecule large enough to be its own heat bath. The population distribution among a limited number of vibrational modes is expected in medium-sized molecules.

Almost all elementary excitation processes (optical excitation, collisions with electrons or atoms, fragmentation) prepare the molecule in an energetically nonequilibrated state with the energy excess contained in a few vibrational modes. The degree and rate of IVR has thus a fundamental importance for chemical and physical processes following electronic or vibrational excitation. If the chemical process such as dissociation or isomerization is more rapid than IVR, the dissociation (isomerization) path depends on the initially excited level. Otherwise, the energy is randomized and the memory of the initial excitation is lost. All statistical theories of monomolecular reactions (such as RRKM) are based on the assumption that its energy is instanta-neously randomized at each step of the reaction path. As will be shown in the further part discussion, this assumption is valid only for a limited class of molecular systems.

3.1.2 The Model of Independent Oscillators

IVR is usually described in terms of the *zero-order* (ZO) *states* $|n\rangle$, eigenstates of the truncated vibrational Hamiltonian with potential

$$\sum_{i=1}^{3N-6} \left(k_i^{(2)} Q_i^2 + k_i^{(3)} Q_i^3 + \cdots \right), \tag{3.2}$$

including higher order diagonal terms but neglecting all off-diagonal terms in the potential and kinetic energy. This approximation is valid for relatively low vibrational levels of rigid molecules; it breaks down for floppy molecules and for high vibrational levels in the vicinity of dissociation or isomerization thresholds. In this approximation, the energies of ZO states are

$$E_n = \hbar \sum_i^{3N-6} \left[\omega_i \left(v_i + \frac{d_i}{2} \right) + x_{ii} \left(v_i + \frac{d_i}{2} \right)^2 + g_{ii} l_i^2 \cdots \right], \tag{3.3}$$

where d_i are degrees of degeneracy, ℓ_i are the angular moments, and their wave functions are products of functions of $3N - 6$ independent oscillators, Q_i,

$$|n\rangle = |\chi_{\mathbf{v}}(Q_1, Q_2, \ldots Q_{3N-6})\rangle = \prod_i^{3N-6} |v_i(Q_i)\rangle, \tag{3.4}$$

where $\mathbf{v}$ represents the set of $3N - 6$ vibrational quantum numbers, v_i. In terms of zero-order states, the optical excitation of an isolated molecule prepares a single bright $|s\rangle$ state (vibrational or ro-vibrational level of the ground or excited electronic state). If assumed that the excitation is uniquely due to transitions from the vibrationless ground-state level, $|0\rangle$, the state is bright when the transition moment μ_{0s} is different from zero, whereas for all other states, discrete $|\ell\rangle$ or forming an $\{m\}$ quasi-continuum, $\mu_{0\ell} = 0$ and $\mu_{0m} = 0$. The selection rules are different for vibrational transitions between the levels of the same electronic state and for vibronic bands in transitions between different electronic states and will be separately defined in Sect. 3.2 and 3.3.

The ZO states are coupled by the $V_{nn'}$ off-diagonal terms resulting from anhar-monic and Coriolis coupling neglected in the zero-order approximation. The fine structure of the absorption spectrum and time evolution of the initially excited state are determined by the $V_{nn'}$ coupling strength and the $\Delta E_{nn'}$-level spacing. One can differentiate three principal cases (Fig. 3.1):

1. If the coupling constant is negligible as compared with the level spacing, $V_{s\ell} \ll \Delta E_{s\ell}$, the s-ℓ coupling may be neglected: the molecule initially excited to the $|s\rangle$ state remains in this state until its deactivation by a photon emission or a collision.

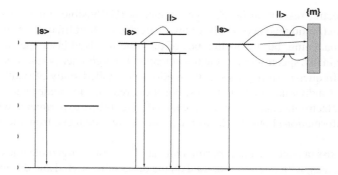

Fig. 3.1. A schematic representation of an $|s\rangle$ level in the absence of the s-ℓ coupling (left), coupled to a discrete set of ℓ levels (center) and (directly and indirectly) coupled to a dense level set $\{m\}$ (right)

2. If $V_{s\ell}$ is of the same order of magnitude as $\Delta E_{s\ell}$ for one or more discrete ℓ levels, the s-ℓ coupling gives rise to a set of discrete $|\varphi\rangle$ levels. The initial excitation of the $|s\rangle$ state (coherent excitation of the set of $|\varphi\rangle$ levels) implies a further evolution, which may be described in terms of reversible transitions between several quasi-resonant ZO states. Such a behaviour corresponding to the *the small molecule case* is called *restricted IVR*.

3. If the density of dark levels effectively coupled to $|s\rangle$ (ρ_{eff}) is so large that they may be approximated by the $\{m\}$ continuum, we attain *the large-molecule case*; the irreversible population transfer from discrete states to a continuum transition is called: *dissipative IVR*.

The stepwise transition from the restricted to dissipative IVR will be discussed later.

The concept of the *effective level density* (ρ_{eff}) is not well defined. The vibrational-level density (ρ_{vib}) for a given molecule with the energy E_{vib} may be estimated when its vibrational spectrum is known by simple level counting or by approximative formulae but, in view of the wide distribution of the coupling constants, $V_{s\ell}$, only a part of them is effectively coupled together. Because ρ_{vib} increases with the size of the molecule and its vibrational energy, we can only anticipate that ρ_{eff} increases in the same way.

Some information about IVR may be deduced from the emission spectrum recorded upon a selective excitation of a single $|s\rangle$ level. If all emission bands may be assigned to transitions from this s-level, one can consider that IVR is absent (or slow as compared with the lifetime of the excited state). The spectrum composed of $s \rightarrow s'$ and $\ell \rightarrow \ell'$ emission bands corresponds to restricted IVR, whereas a complete quenching of the $s \rightarrow s'$ fluorescence indicates the dissipative IVR.

A closer insight into the IVR mechanism may be obtained either by time- and energy-resolved studies of the spectroscopic properties (the spontaneous or stimulated emission, transient absorption, Raman effect) of the system prepared by the pulse excitation or by the high-resolution studies of its absorption spectrum. These two techniques are complementary.

The spectral resolution in the high-frequency (UV/visible) range is limited by large natural (homogeneous) widths of levels $\delta\omega$ due to short lifetimes τ and a strong Doppler broadening, but the techniques of the time-resolved laser spectroscopy are performing in this spectral range. On the other hand, a high-energy resolution attained in the low-frequency (infrared) spectrum allows a detailed study of widths and fine structures of individual rotational lines. For these reasons, the major part of the data on excited electronic states have been obtained in real-time experiments whereas the essential information about the electronic ground state is deduced from the absorption studies.

We discuss in Sect. 3.2 the experimental data for a few model molecules in excited electronic states, with attention focused on the onset of restricted and dissipative IVR and their time scales. In Sect. 3.3, we will limit discussion of the ground electronic state to a restricted class of organic molecules excited to the $v_{XH} = 1$ to 8 levels of the X-H (C-H, O-H) stretching mode. In Sect. 3.4, we will treat IVR as a first step of the vibrational predissociation of small molecules. At last, in Sect. 3.5, we will compare the experimental data with the theory of the level-coupling schemes.

3.2 IVR in Excited Electronic States

3.2.1 Bright and Dark Vibrational Levels

The vibrational structure of the $S_1 \leftarrow S_0$ electronic spectra of polyatomic molecules with a large number of normal modes are composed of a few vibronic transitions: only a small fraction of the zero-order (ZO) levels $|n\rangle$ of the excited state are bright. This limitation is due to the severe selection rules. For allowed electronic transitions between excited (B) and ground (A) electronic state, treated in terms of the Condon approximation, the dipole moment of the $v_A = 0 \rightarrow v_B$ transition is equal to

$$\mu_{0v_B} = \mu_{\text{el}} \prod_{i}^{3N-6} \langle v_i^A = 0 | v_i^B \rangle, \tag{3.5}$$

where the electronic transition moment, μ_{el}, is consider as independent of nuclear coordinates. Two conditions must be fulfilled to get $\mu_{0v_B} \neq 0$:

- The ith normal mode must be totally symmetric; otherwise, only even values of $\Delta v_i = 2, 4, \ldots$ are allowed.
- The potential $U(Q_i)$ must be significantly different in the ground and excited states; when $U_B(Q_i) \equiv U_A(Q_i)$, the vibrational wave functions are orthogonal,

$$\langle v_i^A | v_i^B \rangle = \delta_{v_i^B v_i^A}, \tag{3.6}$$

so that only $v_i^A = 0 \rightarrow v_i^B = 0$ transitions are allowed.

The vibrational modes which fulfill these conditions are called *optically active* $(X, Y, \ldots)$: the $X_0^m Y_0^n$ transitions are allowed and the $|n\rangle = X^m Y^n$ level is a bright $|s\rangle$

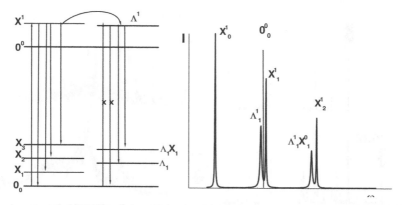

Fig. 3.2. Allowed and forbidden transitions from the 0_0 ground-state level to vibronic levels of the excited state (left), the emission spectra emitted from bright and dark levels (right)

level. All other modes denoted Λ, Π, ... are *optically inactive* and the $\Lambda^\ell \Pi^p$ level is dark as well as the $X^m \Pi^p$ level involving a Π mode for which $\mu_{0p} = 0$ (m, n, ℓ and p indicate the vibrational quantum number in a given mode in the excited, X^m, or ground, X_m, electronic states). The dark states are radiant: the emission from the $\Lambda^\ell X^m$ level to the ground-state $\Lambda_{\ell'} X_{m'}$ levels is allowed provided that $\ell' = \ell$ (Fig. 3.2 left). The emission spectra from nearly isoenergetic levels $|s\rangle(X^m)$ and $|\ell\rangle(\Lambda^\ell)$ are different: the *resonant emission* $s \to s'$ is composed of $X^m_{m'}$ bands with X^m_0 band as origin whereas the Λ^ℓ_ℓ band is the origin of the $\Lambda^\ell_\ell X^0_{m'}$ of the *non-resonant* $\ell \to \ell'$ emission, the $\Lambda^\ell_0 X^0_{m'}$ transitions being forbidden.

As previously quoted, the potentials $U_B(Q_i)$ and $U_A(Q_i)$ for the optically inactive modes are nearly the same so that their ground and excited state frequencies, ω_Π, are only slightly different. The origin of the nonresonant emission is thus close to the 0^0_0 band (Fig. 3.2 right),

$$\omega(\Pi^p_p) = \omega(0^0_0) + p(\omega^B_\Pi - \omega^A_\Pi) \approx \omega(0^0_0), \tag{3.7}$$

and nearly the same for all nonactive $\Pi-$ modes and different p-values. When IVR populates different Π^p and Λ^ℓ levels, their emission spectra overlap and give rise to broad and diffuse bands.

The zero-order states $|s\rangle$, $|\ell\rangle$ and $\{m\}$ as well as the $|\varphi\rangle$ states resulting from their coupling decay with the same rate corresponding to the lifetime of the electronically excited state.

3.2.2 Fluorescence Studies

It seems interesting to follow the IVR story from early studies of the time-integrated emission to the recent real-time experiments.

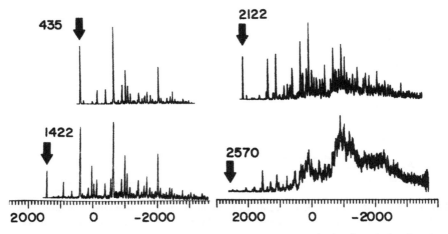

Fig. 3.3. Fluorescence spectra of jet-cooled naphthalene upon the single-vibronic-level excitation as a function of vibrational energy in the S_1 electronic state (from [4])

Time-Integrated Emission Spectra

The emission spectra recorded on the single-vibrational-level (svl) excitation of collision-free molecules in room-temperature vapors showed the absence of IVR for low ($E_{vib} < 1{,}500$ cm^{-1}) S_1 vibrational levels of benzene [1], pyrazine [2] and naphthalene [3]. These studies could be extended to higher vibrational energies only for jet-cooled molecules with $T_{vib} \to 0$ and $T_{rot} \to 0$ so that the overlap of vibronic bands may be suppressed by reduction of the hot band intensities and of the widths of rotational band envelopes. In the pioneering study of naphthalene [4], it was shown that only the resonant $s \to s'$ emission is observed for S_1 levels with $E_{vib} < 2000$ cm^{-1}, whereas, upon the excitation of the $E_{vib} = 2122$ cm^{-1} level, this narrow-band emission is accompanied by a new band system composed of clusters of closely spaced bands with the origin close to the 0_0^0 frequency (Fig. 3.3). These bands are assigned to overlapping $|\ell\rangle \to |\ell'\rangle$ transitions from several dark levels. At higher excitation frequencies (2,570 to 2,867 cm^{-1}), this supplementary emission becomes predominant and, for $E_{vib} \geq 3{,}068$ cm^{-1}, the resonant emission is practically absent. The emission spectrum is composed of diffuse, structureless bands with the intensity distribution similar to that observed in emission of thermally equilibrated naphthalene vapors. The E_{vib} dependence of the emission spectra suggests a transition from the no-IVR limit to the restricted and then to the dissipative IVR when the energies and level densities increase. The dependence of s- to ℓ- intensity ratio, I_ℓ/I_s, on the size and the vibrational energy of the molecule was observed for jet-cooled alkyl-benzenes C_6H_5-$(CH_2)_n$-CH_3 ($n = 0$ to 8) excited to the 0^0, $6b^1 \approx 530$ cm^{-1} and $12^1 \approx 933$ cm^{-1} bright levels of the S_1 state [5]. These molecules have a few optically active ring modes nearly independent of the side chain, whereas the number (density) of low-frequency optically inactive modes of the side chain increases rapidly with its length.

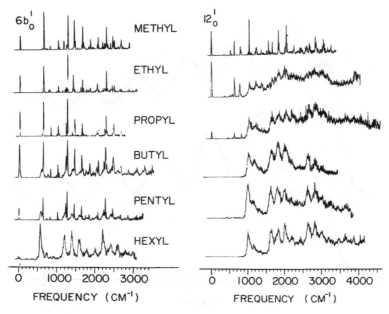

Fig. 3.4. Fluorescence spectra from $6b^1$ and 12^1 levels of the S_1 state (E_{vib} of 520 and 930 cm^{-1}) of alkyl-benzenes (from [5])

The fluorescence spectra emitted from the 0^0 level are nearly the same for all compounds but the emission observed upon excitation of higher levels of the ring modes varies with the chain length. Upon the 12^1 excitation, only the resonant emission is observed for methylbenzene, however, both (resonant and nonresonant) components are present for ethyl and propyl derivatives, whereas, for $n \geq 3$, only the broad-band relaxed emission is found. For the lower $6b^1$ level, the IVR onset is higher: the nonresonant emission appears in butyl-benzene and the resonant one is still present but very weak in the hexyl derivative (Fig. 3.4). A very similar variation of the emission spectra with the energy excess and the length of the side chain was reported for p-alkylanilines [6].

These effects were initially considered in terms of the large-molecule limit as a slow dissipative IVR. In this model, the resonant fluorescence decays with the $k_s + k_{\text{IVR}}$ rate and the $\ell \rightarrow \ell'$ emission component rises with the same rate, k_{IVR} being of the same order of magnitude as k_s. This model was contradicted by measurements of $I_\ell(t)$ and $I_s(t)$ decay curves carried out with an $\sim$10-ns time resolution, which showed that both emission components decay exponentially with the same rate, k_s [7]. It becomes clear that the information deduced from time-integrated emission spectra is limited and must be completed by real-time experiments with a subnanosecond time resolution.

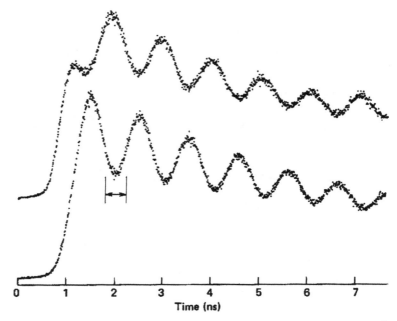

Fig. 3.5. The decay of the resonant ($s \rightarrow s'$) (upper trace) and nonresonant ($\ell \rightarrow \ell'$) (lower trace) components of the fluorescence of anthracene excited to the 1,390 cm^{-1} level of the S_1 electronic state (from [8])

Time-Resolved IVR Studies

In the pioneering studies of IVR in the jet-cooled anthracene excited to the $S_1 - {}^1B_{2u}$ state [8], the time resolution of about 50 ps was attained using an $\sim$1-ps pulse excitation and the time-resolved fluorescence detection. In further experiments [9, 10] using the pump-probe techniques with detection of the excited state population by the two-photon ionization (cf. [11]) or by stimulated emission pumping (TRSEP) and fluorescence depletion (TRFD) [12, 13]), the time resolution was increased to about 100 fs.

All anthracene S_1 levels with $E_{\text{vib}} \leq 1,200$ cm^{-1} show a narrow-band resonance emission and purely exponential decays with the 7×10^7 s^{-1} rate corresponding to the decay rate of the electronically excited state [14, 15]. In contrast, the emission spectrum observed upon the excitation in the 1,380–1,514 cm^{-1} range is composed of the resonant, narrow-band ($s \rightarrow s'$) and nonresonant (relaxed) broad-band ($\ell \rightarrow \ell'$) components. Both components decay with the same $\sim 7 \times 10^7$ s^{-1} rate but their decay is deeply modulated by quantum beats (Fig. 3.5). The beat frequency of the order of 10^9 s^{-1} is the same for s- and ℓ components but they differ by their phases: the $|s\rangle \rightarrow |s'\rangle$ resonant emission does not show any delay with respect to the exciting pulse while the nonresonant emission is delayed by $\Delta\omega/2$ (where $\Delta\omega \approx 1$ ns is the period of the beats). This difference is due to the direct population of the $|s\rangle$ state by the light pulse while the population of the $|\ell\rangle$ state builds up by the $|s\rangle \rightarrow |\ell\rangle$

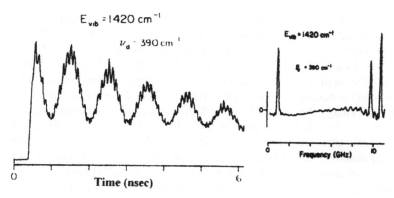

$E_{vib} = 1420 \text{ cm}^{-1}$

$\nu_d \sim 390 \text{ cm}^{-1}$

$E_{vib} \cdot 1420 \text{ cm}^{-1}$

$\xi \cdot 390 \text{ cm}^{-1}$

Frequency (GHz)

Time (nsec)

Fig. 3.6. The decay of the $s \rightarrow s'$ component (the 390 cm^{-1} band) of anthracene fluorescence from the 1,420 cm^{-1} level of the S_1 electronic state and its Fourier transform (from [10])

population transfer and attains its maximum when that of the $|s\rangle$ level goes to zero. Such a time evolution corresponds to the restricted IVR involving a single pair of s and ℓ levels.

The decay of fluorescence from other levels in this energy range shows a modulation by more than one frequency (Fig. 3.6 left). Its Fourier transform (Fig. 3.6 right) is composed of several peaks corresponding to the energy intervals of 0.01–0.03 cm^{-1} between coherently excited $|\varphi\rangle$ levels resulting from the s-ℓ mixing. From frequencies and intensities of these peaks, one can deduce, using the deconvolution technique [16], the energy intervals between the zero-order $|s\rangle$ and $|\ell\rangle$ states and the $V_{s\ell}$ coupling constants. The peaks are narrow, which indicates the absence of an irreversible relaxation of their populations. We are here in the small-molecule strong-coupling case. The fluorescence decay from higher levels has a more complex shape. The resonant fluorescence from the 1,792 cm^{-1}-level is weak and shows an initial rapid decay with the $\sim 4.5 \times 10^{10}$ s^{-1} rate followed by several recurrences and by a long-lived quasi-exponential ($k \approx 1.5 \times 10^8$ s^{-1}) component (Fig. 3.7). Its Fourier transform shows a diffuse structure composed of about 10 broad bands corresponding to a set of overlapping $|\varphi\rangle$ states. The decay of the nonresonant fluorescence may be fitted by a single exponential with the 1.5×10^8 s^{-1} rate weakly modulated by the quantum beats. This picture suggests a sequential $|s\rangle \rightarrow |\ell\rangle| \rightarrow \{m\}$ decay. The limit of the purely dissipative IVR is still not attained.

These results are corroborated by the data reported for other jet-cooled molecules. The onset of the dissipative IVR is much higher than it was previously expected on the basis of experiments carried out in room-temperature vapors (cf. [17, 18]). It seems to correspond to the calculated vibrational level densities of the order of 10^3/cm^{-1}. Because, for the electronically excited molecules with decay times of $\sim$10 ns, the level widths are of the order of $\gamma_m \approx 10^{-3}$ cm^{-1}, this value corresponds with the $\rho_m \gamma_m \approx 1$. Note that the $\rho_m \gamma_m \geq 1$ relation (corresponding to the decay time of m-levels more rapid than the recurrence time) was previously proposed (Sect. 1.4.3) as the onset of the large-molecule limit.

In anthracene, the level density of $\rho_{vib} \approx 120/cm^{-1}$ at 1,792 cm^{-1} corresponds to the sequential IVR. In p-difluorobenzene (p-DFB), IVR is absent at $E_{vib} = 1,650$ and 2,000 cm^{-1} (where $\rho_{vib} = 15$ and $40/cm^{-1}$), is restricted in the 2,070- to 2455-cm^{-1} range ($\rho_{vib} = 50$ and $140/cm^{-1}$) and only for $E_{vib} = 2,885$ cm^{-1} and $\rho_{vib} \approx 500/cm^{-1}$ the rapid ($\tau = 35$ ps) exponential decay suggests the dissipative IVR [11, 19], while in fluorene the onset of dissipative IVR is of the order of 1,700 cm^{-1} [20].

The IVR energy threshold is lower in the presence of low-frequency modes forming a dense set of overtones and combinations. This is the case of low vibrational levels of van der Waals complexes of anthracene and perylene with small polyatomics: the initial population of levels with E_{vib} of 380 and 350 cm^{-1} is rapidly redistributed among the intermolecular vibrations with low (10–30 cm^{-1}) frequencies [21–23]. The IVR rate is also much higher in p-fluorotoluene than in p-difluorobenzene for the same E_{vib}; this difference being due to the coupling between ring modes and low-frequency torsion of the methyl group [24, 25].

The restricted IVR is observed in a wide energy range below the onset of the dissipative IVR. The probability of accidental two- or many-level quasi-resonances increases with the energy excess and level density, but it is impossible to indicate any well-defined threshold. In view of the random distribution of the level spacing and coupling constants, the efficiency of the restricted IVR varies from state to state and from one molecule to the other. For instance, several levels of p-cyclohexylaniline in the $E_{vib} = 480$–820 cm^{-1} range (ρ_{vib} varying from 125 to 1,050/cm^{-1}) show a slow decay of the resonant emission modulated by quantum beats (restricted IVR) while the other ones decay with 50–100 ps decay times (dissipative IVR) [12, 26]. The same conclusion was deduced from the time-integrated fluorescence and fluorescence-excitation spectra of perylene recorded at high resolution in the $E_{vib} = 800$–1600 cm^{-1}, i.e., $\rho_{vib} = 100$ to 2,000 /cm^{-1} range [27]. The difference between the almost isoenergetic levels of two closely related species may also be striking. For instance, the vibrational levels of the gauche form of propylaniline with vibrational energies in the 700–850 cm^{-1} range show a slow decay modulated by well-resolved quantum

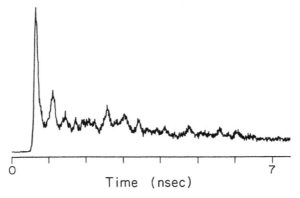

Fig. 3.7. The sequential fluorescence decay from the 1,792 cm^{-1} level of anthracene (from [10])

E $_{vib}$

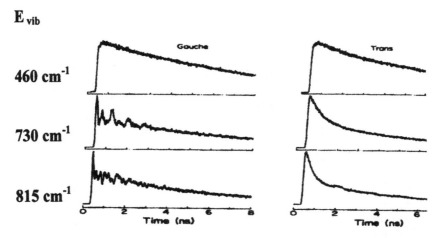

Fig. 3.8. Fluorescence decay from vibronic levels of the S_1 state of trans and gauche forms of p-propylaniline with nearly the same energies (from [28])

beats, whereas the decay of the trans form is quasi-biexponential (Fig. 3.8) [28]. Because the level densities cannot be very different, the level-coupling strength must play an important role.

3.3 IVR in Ground Electronic States

3.3.1 General Remarks

We will treat here several model systems in which the initially excited bright $|s\rangle$ state is the fundamental X_1 or overtone X_m ($m = 2$–8) level of the high-frequency X-H (C-H, O-H, etc.) stretching mode ($\omega_{XH} = 2,800$–$3,600$ cm^{-1}) of such molecules as CHY$_3$ (where Y= D, F, CF$_3$ etc.), C$_2$H$_2$ and its derivatives (R-C≡CH), CH$_3$OH, CH$_3$CN etc. The $|s\rangle$ level is strongly coupled to a limited set of quasi-isoenergetic $|\ell\rangle$ levels of other X-H* stretching modes and to the first overtones of Y-X-H bending modes, all of them being weakly coupled to a dense set of Π_p and combinations $\Pi_p \Lambda_\ell$ of lower frequency modes, Π and Λ. The level-coupling pattern is thus relatively simple as compared to the strongly mixed skeleton modes.

   The difference between bright and dark states is a consequence of the $\Delta v = \pm 1$ selection rule for transitions between vibrational levels of the same electronic state. This rule is strictly valid in the harmonic approximation but is attenuated by the diagonal anharmonicity of large-amplitude X-H vibrations. The anharmonic mixing of v_{XH} levels of the oscillator XH implies a nonzero probability of $\Delta v_{XH} > 1$ transitions, which decreases with increasing Δv but is still nonnegligible for $\Delta v_{XH} = 2$ to 5. In view of the weak anharmonicity and low frequencies ω_Π of skeleton modes Π, the probability of the $v_\Pi = 0 \rightarrow p$ transition to the Π_p level isoenergetic with X_m (where $p \gg m$ because $\omega_{XH} \gg \omega_\Pi$) is negligible.

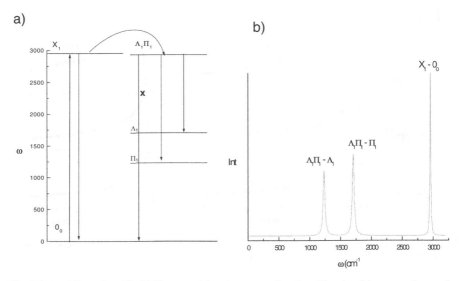

Fig. 3.9. (a) Allowed and forbidden transitions between vibrational levels of the same electronic state and (b) emission spectrum upon the X_1-level excitation

As a matter of fact, the absorption spectra in the near infrared ($5{,}000$–$12{,}500\,\mathrm{cm}^{-1}$) and so-called infrared in the visible spectra in the $12\text{-}500$–$27\text{-}000\ \mathrm{cm}^{-1}$ frequency range of water, ammonia, acetylene etc., composed of well-spaced bands of X-H overtone transitions, have been known for a long time (cf. [29]). The dark states do not appear in absorption spectra; only a few of them are observed in the stimulated emission spectra (SEP) of fluorescent molecules [30, 31]. The energies of other ones are evaluated using the spectroscopic constants determined by the high-resolution infrared and Raman spectroscopy of low ($v = 1, 2, \ldots$) levels. The accuracy of this evaluation is then checked by detection of predicted resonances between bright and dark levels. The maps of dark levels in a wide spectral range were determined in this way for several molecules, e.g., for acetylene in the $E_{\mathrm{vib}} = 0$–$18{,}500\ \mathrm{cm}^{-1}$ range [32–34].

The dark $\Lambda_\ell \Pi_p$ states are radiant, the transitions to lower vibrational states $\Lambda_{\ell-1}\Pi_p$ and $\Lambda_\ell \Pi_{p-1}$ with $\Delta v = -1$ being allowed (Fig. 3.9). Because $\omega_\Pi, \omega_\Lambda \ll \omega_X$, this emission is strongly red shifted with respect to the $X_m \to X_{m-1}$ transition and has a significantly reduced decay rate because of the reduced ω^3 factor. As previously discussed, the long natural lifetimes of vibrationally excited ground-state levels and their small Doppler widths make possible high-resolution absorption studies in the low-frequency spectral region. The 10^{-4}–$10^{-3}\ \mathrm{cm}^{-1}$ resolution is attained in measurements using the narrow-band infrared lasers (color center and Raman-shifted dye lasers, optical parametric oscillators) and linear or skimmered supersonic jets. The real-time experiments with the picosecond time resolution were carried out for molecules in solutions but only a very limited amount of data is available for isolated molecules. We will thus briefly review the results of the real-time fluorescence

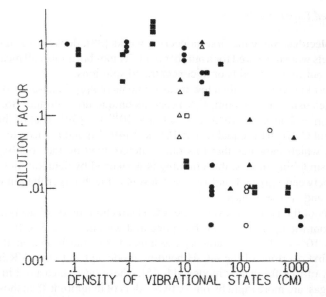

Fig. 3.10. The dilution factor F plotted against ρ_{vib} for $v_{\mathrm{CH}} = 1$ level of different molecules (from [35])

(transient-absorption) studies and discuss in a more detailed way the high-resolution spectroscopy of small- and intermediate-sized molecules.

3.3.2 Emission and Transient Absorption

Time-Integrated Spectra

The first direct observation of IVR from $v_{\mathrm{CH}} = 1$ levels of isolated, jet-cooled molecules was realized by the time-integrated detection of infrared emission spectra upon the nanosecond laser excitation [35]. It was shown that the relative intensity of the resonant $X_1 \rightarrow X_0$ emission decreases with increasing size of the molecule. New bands corresponding to emission from low-energy levels populated by IVR appear but their intensity does not compensate for the decrease of the CH emission. The resulting intensity effects were expressed by a dilution factor, F, the intensity ratio of the observed fluorescence to that expected from absorption measurements ($F = 1$ in the absence of IVR and $F \rightarrow 0$ with increasing IVR rate [35–37]). In a series of molecules, F decreases with ρ_{vib} varying from 10^{-1} to $3 \times 10^3/\mathrm{cm}^{-1}$ and $F < 0.1$ is attained for $\rho_{\mathrm{vib}} \geq 5 \times 10^2/\mathrm{cm}^{-1}$. This level density seems to correspond to the onset of the dissipative IVR from the $v_{\mathrm{CH}}=1$ level (Fig. 3.10).

Time-Resolved Experiments

Isolated Molecules. Since the first pioneering works [38], decay rates of excited vibrational levels were measured for a number of molecules but almost all measurements were carried out in neat liquids or in concentrated solutions.

To our best knowledge, only a few measurements of k_{IVR} in jet-cooled, collision-free molecules using the IR-pump/UV-probe technique are reported. So, the decay rate of the $v_{OH} = 2$-level of HNO_3 is $k = 8.5 \times 10^{10}$ s^{-1} [39]. That of the $v_{OH} = 1$ level of phenol C_6H_5OH is equal to 7×10^{10} s^{-1} [40–43] and reduced to 1.3×10^{10} in C_6D_5OH, which indicates that this rate is determined by the coupling between OH and CH stretching modes: this coupling is weakened by deuteration because the energy gap between $v_{OH} = 1$ and $v_{CD} = 1$ levels is much larger than in the case of the $v_{OH} = 1$ and $v_{CH} = 1$ ones.

The decay of the $v_{CH} = 1$ level of the $\equiv$C-H streching mode of acetylene derivatives in the room-temperature gas is biexponential with the $k_1 = 5 \times 10^{10}$–5×10^{11} s^{-1} and $k_2 = 10^9$–4×10^{10} s^{-1} rates. k_1 is assigned to the transition from the $v_{CH} = 1$ level to the first overtone of the $\equiv$C-H bending mode and k_2 to the IVR involving a large number of low-frequency modes [44, 45]. These rates, measured in the room-temperature gas, are more rapid by one or two orders of magnitude than those deduced from the fine structures of rotational features in the spectra of the same molecules cooled in supersonic beams (see below).

Solvent Effects. In condensed phases, intramolecular processes (IVR) cannot be completely separated from the vibrational relaxation involving the energy transfer to the medium. The data obtained for model systems suggest, however, that the first, rapid steps of relaxation correspond to IVR and are only weakly perturbed by environment effects. The vibrational levels populated by IVR are then depopulated by the energy transfer to the medium with a lower rate of the order of $\sim 10^{10}$ s^{-1}. In the case of acetylene derivatives, the long component of the gas-phase decay is quenched in solutions but the prompt decays with $k_1 \geq 10^{11}$ s^{-1} rates are only slightly accelerated [45]. In the case of benzene and toluene, the $k_1 > 10^{12}$ s^{-1} decay rate is the same in gas and in solution, whereas $k_2 = 2 \times 10^{10}$ and 1.1×10^{11} s^{-1} in the gas are slightly accelerated to $\sim 2.5 \times 10^{11}$ s^{-1} [46]. Similar effects are induced by formation of complexes in supersonic expansions: the decay rate of the $v_{OH} = 1$ level of C_6D_5OH is increased from 1.3×10^{10} in free molecule to 6.7×10^{10} and 1.4×10^{11} s^{-1}, respectively, in its complexes with ethylene and dimethylether [43].

One can thus consider the decay rates measured in solutions as the high limits of the intrinsic k_{IVR} rates, more or less enhanced by intermolecular interactions [47]. The IVR rates estimated in this way for $v_{XH} = 1$ levels of small molecules are contained in the 3×10^{10}–10^{12} s^{-1} limits (cf. [48]).

The time-resolved experiments allow differentiating the ℓ levels that are directly populated from $|s\rangle$: the rise times of their populations are equal to the decay times of that of $|s\rangle$ level: $k_{rise}^{\ell} = k_{rel}^{s}$. The relation $k_{rel}^{s} > k_{rise}^{\ell'}$ indicates that the $|\ell'\rangle$ level is populated by an $|s\rangle \rightarrow |\ell\rangle \rightarrow |\ell'\rangle$ sequential process. For instance, the population of the initially excited $1_1(v_{CH} = 1)$ level of acetonitrile CH_3CN decays with the 5-ps

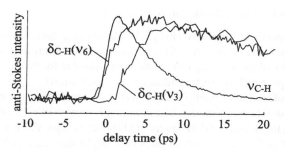

Fig. 3.11. The time dependence of the Raman signals from the CH_3 stretching $\nu_{C\text{-}H}$ and bending $\delta_{C\text{-}H}$ modes of acetonitrile excited to the $\nu_{CH} = 1$ level (from [48])

decay time. The population of the 6_2 level involving the CH_3 bending mode (δ_{CH}) rises during the pulse and continues to rise during the 5-ps decay of 1_1. Conversely, the signal corresponding to the $6_1 3_1 \rightarrow 6_1$ emission has a longer ($\sim$6 ps) rise time, which indicates that the $6_1 3_1$ level is populated indirectly by the $1_1 \rightarrow 6_2 \rightarrow 6_1 3_1$ population transfer (Fig. 3.11) [48].

3.3.3 X-H Absorption Spectra

Introductory Remarks

The progress in the spectroscopy of higher overtones of X-H stretching modes may be illustrated by the history of the C-H overtone bands of benzene, first observed in the early 1930s in the absorption of the neat liquid. In the spectrum of the room-temperature vapor studied by acousto-optic techniques [49], the $40\text{--}100\,\text{cm}^{-1}$ widths of (rotationally unresolved) bands were explained by a homogeneous broadening due to the IVR at the time scale of 50–100 fs. It was shown afterward that these widths are inhomogeneous and result from the Fermi resonance with the C-H bending and from the sequence congestion. The spectra of jet-cooled benzene molecules showed much narrower but still broad bands, the structures of which were fitted by assuming the overlap of a few rotational lines with homogeneous widths of $3\,\text{cm}^{-1}$ for $\nu_{CH} = 2$, of $6\,\text{cm}^{-1}$ for $\nu_{CH} = 3$ and of $17\,\text{cm}^{-1}$ for $\nu_{CH} = 4$ [50–52]. At a higher resolution, the 3-cm^{-1} features of the $\nu_{CH} = 0 \rightarrow 2$ transition split into a number of closely spaced narrow lines [53].

The structure of overtone transitions of X-H modes results from the coupling of the zero-order bright $|s\rangle(\nu_{XH})$ levels to the dark ones. The strength of this interaction varies in wide limits and depends on the overall difference between the vibrational quantum number of both states $\Delta v_{\text{tot}} = \sum_i |v_i^s - v_i^\ell|$. This dependence, a particular case of the energy gap law, discussed in Sect. 1.6, is a consequence of the selection rules for the coupling between vibrational levels of the same electronic state (Sect. 3.5).

The dark levels may be divided into two groups:

- The fundamentals of other stretching XH$'$ modes $v_{XH'}$ and overtones of bending modes of the same or an other XH_n group $|v_b\rangle$. This nonresonant interaction is strong in view of small Δv_{tot} values and is responsible for the coarse structure of absorption spectra (level shifts and intensity distribution).
- A dense set of quasi-resonant $\Pi_p \Lambda_\ell$ levels involving high overtones and combinations of low-frequency modes. The coupling is weaker in view of large values of Δv_{tot}. When the density of dark levels' $|\ell\rangle$ is not very high, this interaction induces a splitting of individual rotational lines of a bright state into a clump of discrete lines corresponding to the set of eigenstates $|\varphi\rangle$. The clump structures vary with J and K and must be studied for individual rotational states. When the density of dark levels is so large that transitions to $|\varphi\rangle$ states cannot be resolved, the clumps look like homogeneously broadened lines. Their apparent line widths are treated in terms of the coupling of a discrete $|s\rangle$ level to the $\{m\}$ quasi-continuum.

X-H Stretch/X-H Bend and X-H Stretch/X-H$'$ Stretch Coupling

In a large class of aliphatic molecules, the frequency ratio between C-H stretching and bending modes, $\omega_s \approx 2900$ cm^{-1} and $\omega_b \approx 1450$ cm^{-1}, is close to 2, which implies the Fermi resonance between pairs of $v_s = n$ and $v_b = 2n$ levels, well known in organic molecules containing -CH$_3$ and/or -CH$_2$-groups [29]. This interaction determines the coarse structure of overtone spectra of such molecules as F$_3$C-H, D$_3$C-H and (CF$_3$)$_3$C-H [55–58]. Their level pattern (when the coupling to the low-frequency modes is neglected) is composed of a set of zero-order states, $|v_s, v_b\rangle$, grouped into polyads of close-lying states characterized by the same quasi-quantum number $N = v_{XH} + v_b/2$. In the C$_{3v}$ symmetry group of X$_3$C-H molecules, where ω_{XH} and ω_b modes are, respectively, of a_1 and e symmetry, the polyad with an even $2N$ value is composed of $N + 1$ A_1 states: $|v_{XH} = N, v_b = 0\rangle$, $|v_{XH} = N - 1, v_b = 2\rangle, \ldots, |v_{XH} = 0, v_b = 2N\rangle$, while for odd $2N$ values, $N + 1/2$ states of E symmetry are $|v_{XH} = N - 1/2, v_b = 1^\ell\rangle$ and so on (where ℓ is the kinetic moment of the degenerate mode). Because the transition moment of the $\Delta v_b > 1$ is negligible, only a single zero-order ($|v_{XH} = N, v_b = 0\rangle$) state of each polyad has a nonzero μ_{0N} transition moment, all others being dark.

The zero-order states of each polyad are coupled together by the off-diagonal anharmonicity in which the main factor is the interaction between $|v_{XH}, v_b\rangle$ and $|v_{XH} \pm 1, v_b \mp 2\rangle$ states.

$$k_{v_{XH}v_b, v_{XH}\pm 1 v_b \mp 2} = k_0[v_{XH}(v_b + 2)^2]^{1/2}/2.$$

The coupling strength depends on v_{XH} and v_b and the level spacing varies with N because of different diagonal anharmonicities of stretching and bending modes (Table 3.1). The level pattern and absorption spectrum of each polyad are thus different (Fig. 3.12).

This scheme may be applied to complex molecules containing several (identical or not) X-H oscillators linked by an X-X bond such as HC-OH (in CH$_3$-OH, CHD$_2$-OH

Table 3.1. Frequencies, coupling and anharmonicity constants for a few X_3CH molecules (in cm^{-1})

Molecule	ω_{XH}	ω_b	k_0	$x_{XH,XH}$	$x_{XH,b}$	$x_{b,b}$
F_3CH	3,080	1,378	±106	-62	-28.5	-6.5
D_3CH	3,047	1,292	25.5	-58	-21.5	-4.5
$(CF_3)_3CH$	3,043	1,356	±65	-57	-14	-0.3

and $HC\equiv C-CH_2-OH$) or $HO-OH$ (in H_2O_2) or $CH-CH$ (in C_6H_6 and its derivatives). For instance, in methanol [45, 59–61], the stretch-bend coupling in the $v_{OH} = 1$ state is not important because of a large frequency difference between the first overtone of the in-plane C-O-H bending (1,340 cm^{-1}) and the O-H stretching mode (3,680 cm^{-1}). Because the anharmonicity is much more pronounced for the ω_{XH} than for the ω_b mode, the quasi-resonance is expected for large N-values. As a matter of fact, the $v_{XH} = 7$ level is split into two components and this splitting is assigned to the $|v_{XH} = 7, v_b = 0\rangle - |v_{XH} = 6, v_b = 2\rangle$ resonance with the ~35 cm^{-1} coupling constant [61]. Still more important for the coarse structure is the interaction between $|v_{OH} = N\rangle$ and $|v_{OH} = N - 1, v_{CH} = 1\rangle$ levels involving O-H and C-H stretching modes. The quasi-resonance is observed for the strongly split $v_{OH} = 5$-level, whereas the $v_{OH} = 4$ and 6 bands have only weak satellites due to off-resonance interactions.

The structure due to the coupling between X-H modes implies an initial periodic time evolution of the molecule upon the coherent excitation by a femtosecond light pulse with beat frequencies of the order of several tenths of femtoseconds. No real-time experiments of this kind have been reported. The fast component observed in the decay of $v_{X-H} \geq 1$ levels was assigned to the energy redistribution within a set of high-frequency modes and this assignment is confirmed by deuterium effect on the decay rate (cf. [48]).

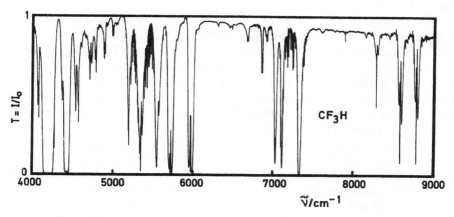

Fig. 3.12. Absorption spectra of CF_3H in the overtone region (from [56])

The Fine Structure of X-H Spectra—ρ_{vib} Dependence

The dependence of the fine structure of the $v_{\text{XH}} = 1$–7 levels (where XH is the acetylenic $\equiv$C-H mode) on the size of the molecule and its vibrational energy excess was studied for jet-cooled acetylene derivatives: propyne [62–67] 1-butyne and 1-pentyne [68], tert-butyl-acetylene, tri-methyl-silyl-acetylene and their deuterated analogues [69–71] and other ones. Unfortunately, no data on emission spectra or decay times of vibrationally cold molecules are available.

In the series: propyne-butyne-pentyne, the overall (calculated) density of vibrational levels ρ_{vib} at the energy of the $v_{\text{XH}} = 1$ level increases from 1 to 80 and 2,400 states/cm^{-1}. As expected, the efficiency of the state mixing increases also in this series. No perturbation has been detected in the $v_{\text{XH}} = 1$ state of propyne (CH$_3$-C$\equiv$C-H), whereas in the $0 \rightarrow 1$ band of 1-butyne (CH$_3$-CH$_2$-C$\equiv$C-H) each rotational transition gives rise to an irregular clump of several (2–10) lines with intervals of 0.005–0.01 cm^{-1}, and $\delta\omega \approx 0.0013$ cm^{-1} Lorentzian widths of individual lines. The total width of the clump is equal to $\delta\omega_{\text{tot}} \approx 0.03$ cm^{-1} (Fig. 3.13 (top)). The rotational lines of the trans and gauche forms of 1-pentyne (C$_3$H$_7$-C$\equiv$C-H) show no fine structure but are strongly broadened; the width of each line, $\delta\omega \approx 0.04$ cm^{-1}, is close to that of the clump in the butyne spectrum. Their profiles are well fitted by assuming that each of them corresponds to a clump of lines (like in butyne) strongly overlapping because their widths, $\delta\omega$, increased to 0.012 cm^{-1} for the trans and to 0.022 cm^{-1} for the gauche form, are larger than their spacing (Fig. 3.13 (bottom)). These results may be rationalized in terms of transition from the strong-coupling to the sequential-decay scheme. In butyne, the coupling of the $|s\rangle$ state to several $|\ell\rangle$ states gives rise to a set of $|\varphi\rangle$ states with small widths limited by the instrumental resolution and the collisional broadening. The total widths of the clumps $\delta\omega_{\text{tot}}$ correspond to the dephasing time of coherently excited state $\tau \approx 200$ ps ($k_{\text{IVR}} \approx 5 \times 10^9$ s^{-1}). The broadening of individual lines in pentyne indicates the coupling of ℓ-states to the $\{m\}$ quasi-continuum and the sequential decay with the rates, $k_{\text{IVR}} = k_{s\ell} \approx 5 \times 10^9$ and $k_{\ell m} \approx 3 \times 10^9$ s^{-1}.

The increase of the mixing efficiency with E_{vib} is obviously due to the E_{vib} dependence of ρ_{vib}. In propyne, the $v_{\text{XH}} = 1$ level is not perturbed and the perturbation is much stronger in the $v_{\text{XH}} = 3$ level than in the $v_{\text{XH}} = 2$ one. The density of totally symmetric A_1 levels, which may be mixed with the v_{XH} state by anharmonicity $\rho_{\text{vib}}(A_1)$, is equal to $\sim$10 for $v_{\text{XH}} = 2$ and to ~ 150/cm^{-1} for $v_{\text{XH}} = 3$. A more efficient level mixing in trifluoropropyne (CF$_3$C$\equiv$C-H) $v_{\text{XH}} = 1$ and 2 states is also due to higher level densities resulting from the presence of lower frequency C-F stretching and bending modes [71].

The level coupling patterns show also a pronounced dependence on the rotational J and K quantum numbers within a single vibrational state. In the propyne $v_{\text{XH}} = 2$ state, the $K = 0$ rotational levels are not perturbed while the $K = 1$–3 ones are split into 2–5 lines with an irregular intensity distribution (Fig. 3.14). In the $v_{\text{XH}} = 3$ state, the clumps assigned to the $K = 0$ subbranch are already composed of a large number of lines but the structure is still more complex for $K = 1$ and cannot be resolved for $K > 1$ subbranches. Such a dependence may be explained by the increase of the

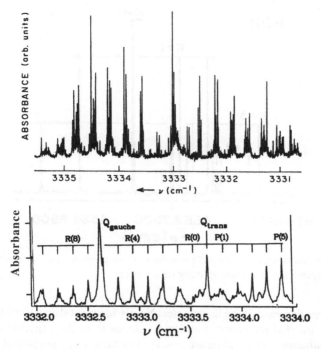

Fig. 3.13. The structure of the $v_{CH} = 0 \rightarrow 1$ band in absorption spectra of butyne (top) and pentyne (bottom) (from [68])

effective level density with J and K due either to the breakdown of the $\Delta K = 0$ selection rule and/or to the increased strength of the Coriolis coupling proportional to K or to $(J^2 - K^2)^{1/2}$ values (cf. Sect. 3.5.1) [64, 65, 68].

The Fine Structure of X-H Spectra—Specific Effects

The simple picture of the IVR dependence on the density of background levels is somewhat misleading and not sufficient to explain all observed effects.

In propyne, the $v_{XH} = 3$ and 6 levels show the fine structure corresponding to the $V_{s\ell} \approx 5 \times 10^{-3}$ cm^{-1} and $\sim 4 \times 10^{-4}$ cm^{-1}, respectively. Surprisingly, the $v_{XH} = 0 \rightarrow 4$ transition does not show any fine structure, whereas the $v_{XH} = 0 \rightarrow 5$ absorption band is completely structureless and diffuse [67].

Striking deviations from the expected ρ_{vib} dependence of decay rates are observed on the Si $\rightarrow$ C substitution and on deuteration of the methyl groups in $(CH_3)_3$X-C $\equiv$ C-H (X = C, Si) molecules [70] (Table 3.2)

The effect of the substitution of C by the Si atom may be tentatively explained by the mass effect on the coupling between vibrations of the $(CH_3)_3$X- and -C$\equiv$C-H groups due to the central X atom [72, 73], but the mass ratio of $\sim$2.5 seems to be not sufficient to induce a decrease of the decay rate by an order of magnitude. The increase of k_{rel} on deuteration is unusual: typically, the deuterium substitution

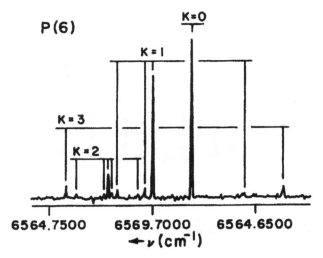

Fig. 3.14. The fine structure of the P(6) transition of the propyne $v_{CH} = 0 \rightarrow 2$ transition. Note the K-dependence of the line splitting (from [64])

induces a large decrease of the decay rates. Nearly identical decay rates for $v_{XH} = 1$ and 2 states are also surprising. Similar deviations from the expected dependence of the level structures and widths are observed in the case of the partial deuteration of the HO-OH molecule [74]. A theoretical treatment of these anomalies would be welcome.

3.4 IVR and the Vibrational Predissociation

The $|s\rangle - v_{XH}$ level of the HXY molecule with energy exceeding the onset of the H-X-Y $\rightarrow$ H-X + Y dissociation is predissociated, i.e., coupled to the dissociative continuum $\{m\}$ directly and through the coupling to a set of $|\ell\rangle$ levels involving low-frequency modes and coupled in turn to $\{m\}$. The same scheme may be applied to the

Table 3.2. Observed v_{XH} level widths $\delta\omega_{tot}$ and decay rates k_{rel} and calculated overall level densities of $(CH_3)_3X$-C≡C-H (X = C, Si) molecules

molecule	$v_{XH} = 1$			$v_{XH} = 2$		
	$\delta\omega$(MHz)	k_{rel} s^{-1}	ρ_{vib}	$\delta\omega$(MHz)	k_{rel} s^{-1}	ρ_{vib}
X = C						
h_9	800	5×10^9	4.9×10^2	1400	8.5×10^9	6.2×10^5
d_9	4000	2.5×10^{10}	2.8×10^3	~8000	$> 5 \times 10^{10}$	7.6×10^6
X = Si						
h_9	75	0.5×10^9	10^4	~40	0.25×10^9	2.9×10^7
d_9	175	1.2×10^9	10^5	1100	7×10^9	6.0×10^8

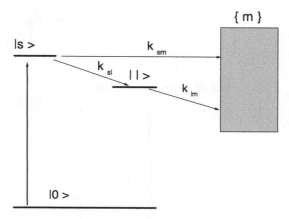

Fig. 3.15. Rate constants of the direct and indirect $s \leftrightarrow m$ coupling

dissociation of weakly bound molecular complexes (discussed in Sect. 4) and to the autoionized levels above the ionization limit. [75]

IVR and Dissociation Rates

When the $|s\rangle$ state is coupled to a single ℓ state (or of a set of equivalent ℓ states), all of them being coupled to the $\{m\}$ dissociative continuum, the irreversible evolution of the excited system is expressed by three rate constants: k_{sm}, $k_{s\ell}$ and $k_{\ell m}$ corresponding, respectively, to the direct s dissociation, to IVR, and to the ℓ-m transition (Fig. 3.15).

One can differentiate three limiting cases:

- When the direct dissociation channel is predominant ($k_{sm} \gg k_{s\ell}$), the decay rate of the s-level ($k_s = \gamma_s/\hbar$) and that of the rise of the fragment concentration are identical: $k_{\text{rise}} = k_s \approx k_{sm}$. One can expect a monotonic increase of this rate with increasing energy excess above the dissociation onset: $\Delta E = E_{\text{vib}} - D_0$. The s level is broadened and structureless.

- When the direct s-m coupling is weak ($k_{sm} \to 0$) but $k_{\ell m} \gg k_{s\ell}$, IVR is the slow, rate-determining step. As previously, the fine structure of the initially excited level cannot be resolved and $k_{\text{rise}} = k_s$ but now $k_s \approx k_{s\ell}$. Because the s-ℓ coupling patterns vary from one s level to another, a pronounced and irregular J and v dependence of the dissociation rates would not be surprising.

- At last, in the limit, $k_{s\ell} \gg k_{sm}$ and $k_{s\ell} \gg k_{\ell m}$, the fast IVR and slow dissociation implies a well-resolved structure of the $0 \to s$ absorption line with an overall width $\delta\omega_{\text{tot}} \approx \hbar k_{s\ell}$. The growth of the fragment concentration (k_{rise}) is slower than the decay of the s state, the rate-limited step being dissociation of ℓ levels.

The values of the constants may be deduced from the level widths and from the pump-probe experiment. For instance, in the case of HOCl and HO-OH molecules upon a selective excitation of a single rotational feature by one or two photons, the dissociation product (OH) is monitored by LIF: the fluorescence induced by a UV

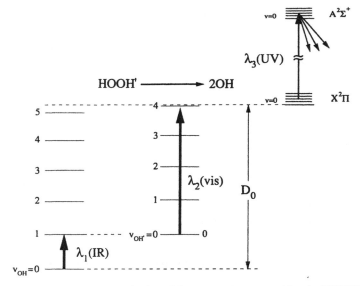

Fig. 3.16. Scheme of a two-laser excitation of the $v_{OH} = 4$, $v_{OH'} = 1$ level of HOOH followed by LIF detection of free OH fragments (from [76])

probe laser (Fig. 3.16). The fine structure and widths of absorption lines, γ, are determined by scanning the pump laser with a constant delay between the pump and probe pulses. On the other hand, by varying the probe-pump delay, one can determine the rate of formation of the dissociation products, i.e., the dissociation rate.

Unfortunately, complete information about the level widths and decay times is still not available, even in the case of the best known model molecules, HOCl and hydrogen peroxide H_2O_2 [74, 76–82].

HOCl

The $v_{OH} = 7$ (21,709 cm^{-1}) level has a large energy excess with respect to the onset of the HOCl $\rightarrow$ OH + Cl dissociation at 19,290 cm^{-1} but the $J = 0$, $K = 0$ level of the $v_{OH} = 6$ state is situated at 19,124 cm^{-1} so that only its high rotational levels ($J \geq 18$ for $K = 0$ and $J \geq 10$ for $K = 3$) are predissociated. A strong perturbation by resonance with the $|v_{OH} = 4$, $v_{ClO} = 4$, $v_{bend} = 2\rangle$ state at 19,137 cm^{-1} gives rise to a Fermi doublet. The widths of rotational lines of both doublet components with the energy excess of 1–350 cm^{-1} above the dissociation onset are limited by the instrumental widths of 0.03 cm^{-1} so that only the high limit of the *s*-level decay rate, $(k_{s\ell} + k_{sm}) \leq 5 \times 10^9$ s^{-1}, is established. The rise times, k_{rise} of the free OH signal vary in extremely wide limits of 10^6–3×10^8 s^{-1} within the band and seem to be independent of the energy excess. They decrease with increasing K in spite of increasing level densities and seem to fluctuate randomly with J for constant K values. The ratios of k_{rise} constants of two (J, K) levels belonging to two components of the Fermi doublets vary also in an irregular way (Fig. 3.17). Because the decay

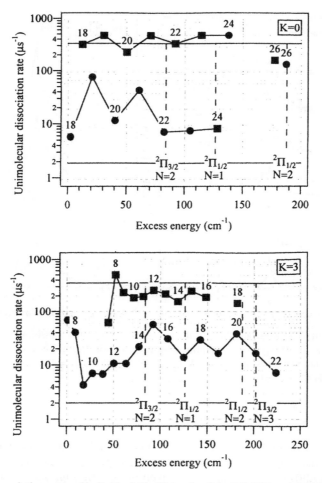

Fig. 3.17. Dissociation rates of individual (J, K) levels of the HOCl Fermi doublet $(6, 0, 0) +$ $(4, 2, 2)$ (from [78])

time of the s level could not be measured, we have no clear indication whether the dissociation mechanism is direct or sequential. It seems, however, difficult to believe that the dissociation rates (k_{sm} or $k_{\ell m}$) vary strongly with J and K and are insensitive to the energy excess. The wide distribution of the decay rates may be rather assigned to the J, K dependence of the k_{IVR} ($k_{s\ell}$) rates resulting from a random variation of the spacing $\Delta E_{s\ell}$ between bright rotational levels (J, K) of the v_{OH} state and dark $(J' = J$ and $K' = K)$ levels. The treatment of these perturbations using the tier model (discussed in Sect. 3.5) allows one to reproduce qualitatively the observed phenomena [78].

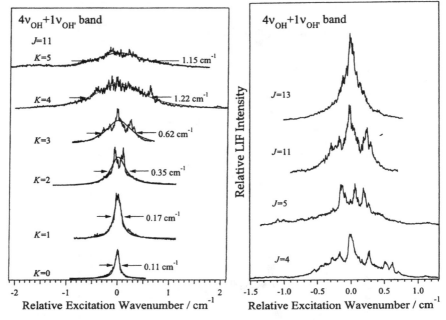

Fig. 3.18. (Left) The K dependence of the contours of rotational transitions to the $J = 11$ levels of the $v_{OH} = 4, v_{OH'} = 1$ vibrational state of HOOH. (Right) The J dependence of the contours of rotational transitions to the $K = 3$ levels of the $v_{OH} = 4, v_{OH'} = 1$ vibrational state of HOOH (from [76])

H_2O_2

The overtone and combination levels of two nearly independent stretching OH and OH′ modes: the $|v_{OH} = 7\rangle$, $|v_{OH} = 6\rangle$ and $|v_{OH} = 3, v_{OH'} = 2\rangle$ levels with 21,422, 18,948 and 17,307 cm^{-1} energies, respectively, have their origins above the $H_2O_2 \rightarrow$ 2OH dissociation onset of 17,050 cm^{-1}. The origin of the $|v_{OH} = 4, v_{OH'} = 1\rangle$ state is situated slightly below it (16,937 cm^{-1}) so that the energy of the $K = 3, J = 4$ rotational level exceeds the onset by 4 cm^{-1}. No real-time experiments have been yet reported for H_2O_2, but their spectra contain a rich information about the vibrational predissociation mechanism. The lowest predissociated level ($K = 3, J = 4$ level of the $v_{OH} = 4, v_{OH'} = 1$ state) appears as a clump of partially resolved lines with the overall width $\delta\omega_{tot} = 0.71$ cm^{-1}, whereas the widths of individual lines are $\delta\omega \geq 0.08$ cm^{-1}. This picture corresponds to $k_{s\ell} = 1.35 \times 10^{11}$ s^{-1} and $k_{\ell m} \approx 1.5 \times 10^{10}$ s^{-1}, i.e., to the sequential decay with $k_{s\ell} > k_{\ell m}$. At higher ΔE_{vib}, the $k_{s\ell}/k_{\ell m}$ ratio is reduced so that the fine structure of the clump cannot be resolved. Total widths of clumps corresponding to the same ($K = 2, J = 11$) level of different vibrational states increase slowly with the energy excess ΔE above the dissociation onset: from $\delta\omega_{tot} = 0.35$ cm^{-1} for $\Delta E = 59$ cm^{-1} to $\delta\omega_{tot} = 0.95$ cm^{-1} for $\Delta E = 400$ cm^{-1} and to $\delta\omega_{tot} = 1.92$ cm^{-1} for $\Delta E = 2,433$ cm^{-1}. Their profiles may be fitted by assuming an increase of the dissociation rate, $k_{\ell m}$, and a practically constant IVR rate, $k_{s\ell}$, so that the process is still sequential but with $k_{s\ell} \approx k_{\ell m}$. A sudden jump of $\delta\omega_{\text{tot}}$ to ~ 30 cm^{-1} for the $|v_{\text{OH}} = 7\rangle$ state ($\Delta E = 4{,}512$ cm^{-1}) seems to be due to the opening of a new dissociation channel (formation of the OH fragment in its excited $v = 1$ state).

The fine structure of absorption spectra varies with J and K quantum numbers in a still not elucidated way. As shown in Fig. 3.18 left, the $K = 0$, $J = 11$ level of the $|v_{\text{OH}} = 4, v_{\text{OH}'} = 1\rangle$ state shows no fine structure and has a Lorentzian shape with the 0.11 cm^{-1} width. The overall width of the level increases with K, but it splits for $K = 2$ and 3 into several narrow component lines. For the $K > 1$ levels with energies close to the dissociation threshold, the width of the clump of lines is reduced for higher J values until all lines coalesce into a single quasi-Lorentzian line (Fig. 3.18 right). The mechanism of such a motional narrowing with increasing frequency of rotation is not obvious.

3.5 Modelling of the Level Coupling Patterns

The IVR processes are induced by the coupling between zero-order states $|n\rangle$ assumed as eigenstates of the Hamiltonian H_0 of the harmonic approximation (cf. Sect. 3.1) neglecting the interaction between vibration and rotation. This coupling is due to two types of cross-terms in the exact Hamiltonian involving the off-diagonal anharmonicity of vibrations and the vibration–rotation interaction: Coriolis and centrifugal coupling.

3.5.1 Level Coupling Selection Rules

Anharmonic Coupling

The harmonic approximation of the potential energy surface by the function

$$U(Q_1, Q_2, \ldots, Q_{3N-6}) = \sum_i^{3N-6} k_i Q_i^2 / 2 \tag{3.8}$$

(where Q are normal coordinates) is valid only for infinitesimal vibrations. For final but still small amplitudes, the deviation from the $k_i Q_i^2/2 + k_j Q_j^2/2$ paraboloid for each pair of coordinates is expressed by cross-terms,

$$U_{ij} = k_{ijj} Q_i^2 Q_j + k_{ijj} Q_i Q_j^2 + \ell_{iiij} Q_i^3 Q_j + \ell_{iijj} Q_i^2 Q_j^2 + \ell_{ijjj} Q_i Q_j^3 + \cdots . \tag{3.9}$$

If the off-diagonal terms appearing in the kinetic energy T_{ij} of the molecule are neglected, the vibrational Hamiltonian may be approximated by

$$H = H_0 + \sum_{i,j,k} \Phi_{ijk} Q_i Q_j Q_k + \sum_{i,j,k,l} \Phi_{ijkl} Q_i Q_j Q_k Q_l + \cdots , \tag{3.10}$$

with the Φ coefficients decreasing rapidly with the term order. The coupling matrix element between the $|m\rangle$ and $|n\rangle$ states differing by quantum numbers in ith and jth mode, $|m\rangle = |v_i\rangle|v_j\rangle|\Pi_v\rangle$ and $|n\rangle = |v_i'\rangle|v_j'\rangle|\Pi_v'\rangle$ (where $|\Pi_v\rangle$ and $|\Pi_v'\rangle$ are products of vibrational wave functions for all other modes) is a sum of terms of the type

$$H_{ij}' = \langle m|\Phi_{ijj}Q_i Q_j^2|n\rangle = \Phi_{ijj}\langle v_i|Q_i|v_i'\rangle\langle v_j|Q_j^2|v_j'\rangle\langle\Pi_v|\Pi_v'\rangle, \tag{3.11}$$

where the matrix elements $\langle v_i|Q_i|v_i'\rangle$ are given by

$$\langle v_i|Q_i|v_i'\rangle = \sqrt{\frac{h}{8\pi^2 c\omega_i}}\sqrt{v\pm 1}. \tag{3.12}$$

The Q_i are assumed to represent normal coordinates involving mass-weighted displacements, ω_i is the harmonic frequency in cm^{-1}. If the dimensions of mass are in amu and displacements in a_0, the factor $h/8\pi^2 c$ amounts to 60.1997 amu a_0^2 cm^{-1}. The matrix elements are obtained from Eq. 3.12 by matrix multiplication. We obtain selection rules for the third-order coupling between two modes as

$$v_i' = v_i \pm 1, \quad v_j' = v_j \pm 2, \quad \Pi_v' \equiv \Pi_v. \tag{3.13}$$

The overall change of vibrational quantum numbers,

$$\Delta v_{\text{tot}} = \sum_i |\Delta v_i|, \tag{3.14}$$

is thus equal to the order of the anharmonicity term. The fourth-order term may couple the states with $\Delta v_{\text{tot}} = 4$, but its efficiency is strongly reduced because of much lower values of Φ_{ijkl} constants. A direct coupling of states with $\Delta v_{\text{tot}} > 4$ is still less efficient and may be neglected, except for very large vibration amplitudes.

As long as the symmetry of the molecule is not perturbed by large-amplitude nontotally symmetric vibrations, the anharmonicity may mix only the states belonging to the same representation of the symmetry group. The same rule is valid for a molecule deformed by nontotally symmetric vibrations within its reduced symmetry group. The anharmonicity mixes the levels with the same $J = J'$ and $K = K'$ rotational quantum numbers, the mixing strength being independent of J and K.

Coriolis Coupling

The effects due to rotation-vibration coupling (centrifugal distortion and Coriolis effect) [83] depend on rotational quantum numbers and tend to zero for $J \to 0$. The first of them is weak and usually neglected; it may be, however, nonnegligible for high rotational levels being proportional to J^2 and not to K or $\sqrt{J^2 - K^2}$ as the Coriolis effect.

For simplicity sake, we describe the Coriolis effect in the simple case of symmetric top molecules, such as CXY_3 with the C_{3v}. symmetry. The perturbation strength is given by [83, 84]

$$H' = \sum_\sigma p_\sigma J_\sigma / I_\sigma \qquad (\sigma = x, y, z) \qquad (3.15)$$

where J_σ p_σ are projections of the total and vibrational kinetic moments on molecular axes and I_σ the corresponding moments of inertia. The vibrational angular momentum components are related to the $3N - 6$ coordinates Q and associated momenta P according to

$$p_\sigma = \sum_{r,s=1}^{3N-6} \zeta_{rs}^\sigma Q_r P_s, \qquad (3.16)$$

where ζ are the Coriolis coupling constant, which in turn depend on the masses, the nuclear geometry and the potential energy function.

The matrix element of H' between two ro-vibronic states $|m\rangle = |v_i, v_j, J, K\rangle$ and $|n\rangle = |v_i', v_j', J', K'\rangle$, may be factorized as

$$
\begin{aligned}
H'_{mn} &= \langle v_i, v_j, J, K | H' | v_i', v_j', J', K' \rangle \\
&= \frac{\langle v_i, v_j | p_z | v_i', v_j' \rangle \langle J, K | J_z | J', K' \rangle}{I_a} + \frac{\langle v_i, v_j | p_\pm | v_i', v_j' \rangle \langle J, K | J_\pm | J', K' \rangle}{I_b}
\end{aligned}
$$

wherefrom the selection rules for rotation and vibration are deduced. The nonzero matrix elements for rotation around the top axis, z, are

$$\langle J, K | J_z | J, K \rangle = \hbar K \qquad (3.17)$$

and for rotation around the axes perpendicular to z,

$$\langle J, K | J_\pm | J, K \pm 1 \rangle = \hbar [J(J+1) - K(K \pm 1)]^{1/2} \qquad (3.18)$$

The coupling strength is thus proportional to K or to $\sqrt{J^2 - K^2}$.

In order to conserve the overall (rotational $\otimes$ vibrational) symmetry, the vibrational symmetry of the initial and final state must be the same when $K' = K$ and different when K and $K' = K \pm 1$ states differ by their rotational symmetry. In the C_{3v} symmetry group, p_z and $p_\pm$ belong to the a_2 and e representations so that p_z may couple two degenerate (e) states whereas $p_\pm$ has nonzero matrix elements between a_1 and e levels. On the other hand, Δv selection rules indicate that the only nonzero matrix elements are

$$\langle v_i, v_j | p_z | v_i \pm 1, v_j \mp 1 \rangle \qquad (3.19)$$
$$\langle v_i, v_j | p_\pm | v_i \pm 1, v_j \pm 1 \rangle. \qquad (3.20)$$

As in the case of the anharmonic coupling, the Coriolis effect allows the direct coupling of a small number of states.

The Coriolis mixing of K and $K \pm 1$ states implies that the (v, J, K) state interacts not only with $K' = K$ but with different K sublevels, v', $J' = J$, K'. This interaction is enhanced in the molecules in which K is not an entirely good quantum number, for not exactly symmetric-top molecules. The effective density of $|\ell\rangle$ states coupled to $|s\rangle$ $\rho_{eff} \leq \rho_{vib}$ in absence of rotation-vibration interaction may thus increase to $\rho_{eff} \leq (J + 1)\rho_{vib}$ for high J values.

The J dependence of the effective level density suggests a pronounced dependence of the IVR rate on the average $\langle J \rangle$ value, i.e., on the temperature. To our best knowledge, this dependence was never systematically investigated. The nonconservation of the K value in an isolated molecule allows the vibration-rotation energy exchange. Because the projection of the molecular transition moment on the laboratory axes (i.e., polarization of fluorescence) depends on K, the existence of $\Delta K \neq 0$ transitions implies the depolarization of the resonance fluorescence of collision-free molecules. The polarization measurements suggest that this process takes place in several rotationally hot medium-sized molecules [85, 86].

3.5.2 The Level Pattern and the Tier Model

The anharmonic as well as the Coriolis mechanisms induce an efficient direct coupling only between the states differing by small Δv_{tot}, all others being very weakly coupled by higher order anharmonicity terms. A more efficient mixing is due to a sequential coupling within a chain of states, $s \leftrightarrow \ell \leftrightarrow \ell' \cdots \ell''$. The s-ℓ'' interaction may be different from zero (even for large Δv_{tot} value) when the coupling strength is large enough in each $\ell \leftrightarrow \ell'$ intermediary step, but its overall strength decreases with the number of steps.

The Δv_{tot} dependence of the IVR is thus a direct consequence of the selection rules. We are in the sequential coupling case and the overall vibrational level density does not play any role. A more realistic treatment of IVR is the *tier model* based on the concept of level coupling chains [87–89]. This model takes into account all the $V_{s\ell}$ and $V_{\ell\ell'}$ terms evaluated by the low-order perturbation treatment for all states contained in a broad ΔE energy band provided that the values of the $V_{s\ell}/\Delta E_{s\ell}$ and $V_{\ell\ell'}/\Delta E_{\ell\ell'}$ ratios are nonnegligible. The terms due to higher order perturbations are neglected. It is thus possible to establish the chains of mutually coupled states: a few of them strongly coupled to the s-state form the first tier, the second one being composed of states strongly coupled to the first tier states and so one (Fig. 3.19). In this scheme, the transition from the restricted to the dissipative IVR is represented by a very simple picture. When the chains of states are not interrupted, i.e., when an important fraction of the nth tier states are coupled to more than one state of the $(n + 1)$ tier, the s state is efficiently coupled to a large number of quasi-resonant states (dissipative IVR). On the other hand, if this chain is interrupted between the n and $n + 1$ tiers (i.e., when all n to $n + 1$ coupling terms are negligibly small), IVR is restricted to a subspace of a few modes. Between these limits, a strong coupling of $|s\rangle$ to a few $|\ell\rangle$ states of the first tier gives rise to a few well-separated levels with fine structure (resolved or unresolved) resulting from the weaker coupling to a dense set of further tier levels. The tier model was applied to the simulation of the fine structure of fundamental and overtone levels of the C-H stretching mode of n-propyne [88, 89]. This treatment underestimates the number of fine structure components but reproduces its essential parameters and gives a correct semiquantitative description of the level coupling scheme.

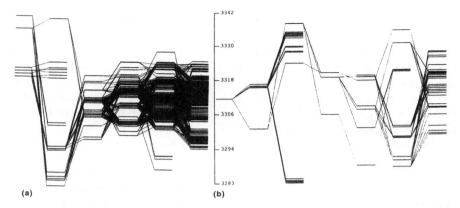

Fig. 3.19. Scheme of first tiers of sequentially coupled vibrational levels: (a) continuous chains corresponding to dissipative IVR, (b) interrupted chains, restricted IVR (from [87])

3.6 Conclusions

The IVR processes in excited electronic states have been studied by time-resolved techniques, whereas the major part of information about the ground electronic states is deduced from the high-resolution absorption spectroscopic studies. Experimental data obtained for the same molecule in different electronic states are still nonavailable but there is no indication of a significant difference between IVR mechanisms or rates in the S_1 and S_0 electronic states.

The strength of the coupling between the zero-order states varies in extremely wide limits so that the concepts of the average coupling strength and average level density are inadequate. Typically, the bright s level is strongly coupled to a few ℓ dark levels, whereas the coupling to a dense set of m levels is much weaker. The striking example is that of ground-state C-H modes, where the coupling constant is of the order of 50 cm^{-1} for $(v_s = n) \leftrightarrow (v_b = 2n)$ stretch-bend interaction and of the order of 10^{-2}–10^{-1} cm^{-1} for interaction with overtones of low-frequency modes.

Such a coupling pattern implies a two-step (or many-step) time evolution composed of an energy redistribution between X-H modes taking place at the femtosecond time scale followed by a slow energy flow to other modes with ~10-ps characteristic times.

The evolution of the IVR mechanisms with the size of the molecule and its vibrational energy content is well described within the scheme of the theory of nonradiative processes as transition from the small-molecule limit through the strong-coupling and sequential-coupling cases to the large-molecule limit.

Because the $V_{s\ell}$ coupling constants are small for most of the dark levels, the probability of a $\Delta E_{s\ell} \approx V_{s\ell}$ quasi-resonance is negligible even in relatively large molecules with vibrational energies E_{vib} of the order of the average vibrational frequency $\langle \omega_{\text{vib}} \rangle$. At higher E_{vib}, the probability of two-level or a few-level resonances increases and restricted IVR is observed for a large fraction of s levels in a wide energy range. This process is well described in terms of the weak- or strong-coupling case, as evidenced by fine structure of absorption spectra composed of narrow lines and by the quantum-beat modulation of fluorescence. The threshold of dissipative IVR is much higher and corresponds to the effective density of vibrational levels ρ_{eff} so large that the levels of a dense $\{m\}$ set overlap, $\langle \Delta E_{mm'} \rangle = 1/\rho_m \leq \gamma_m$. Even in this energy range, the IVR mechanism corresponds rather to the sequential $s \to \ell \to \{m\}$ than to the direct (large-molecule limit) $s \to \{m\}$ transition. The rates of dissipative IVR are, in the ground as well in the excited electronic states of intermediate-sized molecules, contained in the $k_{\text{IVR}} = 10^{10}$–10^{12} s^{-1} range.

The pump-probe experiments show that, with picosecond- or subpicosecond pulses, one can induce either the $s \to s'$ or $\ell \to \ell'$ transition by varying the delay of the probe with respect to the pump pulse. With a good choice of molecules and frequencies, it is thus possible to attain two different higher excited states and induce two different photo-dissociation (or photo-isomerization) processes. The first attempts of the bond selective photochemistry (dissociation induced by sequential many-photon pumping with the CO_2 laser radiation [90]) failed because of the slow pumping rate as compared with the IVR rate. Such a project is now more realistic in view of progress in short-pulse laser techniques. As a matter of fact, the preliminary data indicate that the yields of competing reactions induced by the femtosecond pulse pumping depend on the pulse shapes [91].

4

Vibrational and Rotational Relaxation

4.1 Introduction

The vibrational and rotational relaxation in the gas and condensed phases is a multistep process, but we are interested only in an elementary step: a transition between two energy levels of the molecule induced by interaction with its environments. The simplest model case is that of a diatomic molecule XY colliding with a rare gas atom C, forming an XY.C complex or isolated in an infinitely extended rare-gas crystal (matrix) C_∞. Several conclusions from the study of these model systems may then be applied to the case of polyatomics (the XY symbol is used also in this case) and to many-body interactions in condensed phases.

We will treat here transitions between the levels of the same (ground or excited) electronic state, whereas those involving the vibrational levels of different electronic states will be discussed separately in Chapter 5.

4.1.1 Intermolecular Potentials

The configuration of a system composed of two or more molecules considered as rigid bodies is described by a set of *external coordinates* Q. The number of independent coordinates varies from $n = 2$ for XY + C to $n = 6$ for two polyatomic nonlinear molecules or for the molecule trapped in a rigid C_∞ lattice.

If XY is rigid with the interatomic distance r_{XY} = constant, the potential energy of the XY + C system is a function of two variables R and θ (Fig. 4.1 (a)): $U(R, \theta)$. Obviously, $U(R) \to +\infty$ for $R \to 0$ and $U(R) = 0$ for $R \gg R_{int}$, where the interaction limit R_{int} exceeds by a factor of 10 the sum of van der Waals radii d_{XY} and d_C and is of the order of 2×10^{-7} cm^{-1}. The depth of the energy minimum $U(R^{eq}, \theta^{eq})$ at the equilibrium configuration of the complex XY·C corresponds to its bonding energy D_e.

In reality, XY is not rigid, but its vibrational frequency is so large that only the average $\langle r_{XY} \rangle$ value determines the strength of XY.C interactions. Because $\langle r_{XY} \rangle$ depends on the v_{XY} vibrational quantum number, the intermolecular potential is more

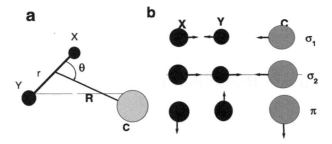

Fig. 4.1. (a) Coordinates of the XY.C complex and (b) normal modes of XY.C in a linear configuration

or less modified by the vibrational excitation of XY. The potentials are currently evaluated for individual molecular systems by *ab initio* or semiempirical techniques. Their validity is checked by comparing the predicted and observed structures and bonding energies of XY.C complexes. Nevertheless, the intermolecular potentials are still often characterized by simple models [1]. So, the R-dependence is approximated by one of the formulae, such as Morse,

$$U(R) = D_e(1 - \exp[-\beta(R - R^{eq})])^2; \tag{4.1}$$

Lennard-Jones (LJ),

$$U(r) = 4\varepsilon[(R_0/R)^{12} - (R_0/R)^6]; \tag{4.2}$$

or 6-exp:

$$U(r) = Ae^{\delta R} - \varepsilon[(R_0/R)^6]; \tag{4.3}$$

and the simplest one, a purely repulsive hard-sphere (HS) potential,

$$U(R < R_0) = +\infty, \quad U(R \geq R_0) = 0. \tag{4.4}$$

The D_e, R^{eq}, R_0, ε, ... parameters are chosen to fit the experimental data. The angular $U(\theta)$ dependence is usually approximated in terms of spherical harmonics. The energy minima of XY·C systems as a rule correspond either to the linear configuration ($C_{\infty v}$ symmetry) or to the T-shaped configuration (the C_{2v} symmetry in the case of a homonuclear X_2 molecule).

The essential effect of the XY.C interaction is the coupling between vibrational, rotational and translational degrees of freedom. At $R \gg R_{int}$ the structure of the free XY + C pair is described by nine independent coordinates, the energy corresponding to each of them being conserved:

- three translations of the XY.C center of mass,
- three translations of XY and C with respect to their center of mass Tr_x, Tr_y and Tr_z,
- two rotations of XY around the x and y axes Rot_x, Rot_y,
- the X-Y (internal) stretching vibration,

with constant components of E_{cm}, E_{trans}, E_{rot}, and E_{vib}.

At $R \leq R_{int}$, different degrees of freedom are coupled together and the energy of the system is diagonalized in the basis of new coordinates—combinations of the previous ones. We will illustrate this treatment by the simplest case of a linear XY-C complex assuming that infinitesimal deviations from this configuration may be approximated by vibrations in the harmonic potential $U(R, \theta) = f_R(R - R^{eq})^2 + f_\theta(\theta - \theta^{eq})^2$. Three of the nine degrees of freedom of XY.C correspond now to translations of its center of mass, two others to rotations of the whole system and the remaining four degrees of freedom give rise in the $C_{\infty v}$ symmetry group to two σ stretching modes and to a degenerate bending π mode (Fig. 4.1 (b)).

The normal coordinates of the stretching modes σ_1 and σ_2 are linear combinations of *internal* X-Y stretching and *external* XY-C stretching corresponding to coordinates $q = r - r^{eq}$ and $Q = R - R^{eq}$ equivalent to the τ_z translation,

$$\sigma_1 = \alpha q_{XY} + \beta Q_{Y.C}$$
$$\sigma_2 = -\beta q_{XY} + \alpha Q_{Y.C},$$

with frequencies $\omega_1 \approx \omega_{XY}$ and $\omega_2 \approx \omega_{XY.C}$. Because the frequency of the internal mode ω_{XY} is much higher than that of the external $\omega_{Y.C}$ mode, the $Q - q$ mixing is weak ($\alpha \gg \beta$).

The components of the π mode are linear combinations of rotational and translational coordinates,

$$\pi_x = \alpha T r_x - \beta Rot_y$$
$$\pi_y = \alpha T r_y - \beta Rot_x$$

(the combinations with sign "+" correspond to the overall rotation). The mixing of low-frequency rotational and translational movements is thus strong ($\alpha \approx \beta$).

One can easily show that, in the X_2C molecule with a T-shape (C_{2v}) configuration and harmonic potential, the Tr_z translation is mixed with the internal vibration yielding two a_1 modes, whereas the Tr_x translation and Rot_y give rise to the bending b_2 mode. The mixing is more extended in an anharmonic potential corresponding to larger deviations from the symmetric equilibrium configuration.

A *substitution center* XY.C_∞ in a crystal may be treated in a similar way: the molecule XY occupies in the site XYC_n of the C_∞ lattice a position corresponding to the energy minimum so that each displacement ΔR or $\Delta \theta$ from this configuration gives rise to the restoring force $(\partial U/\partial R)\Delta R$ or $(\partial U/\partial \theta)\Delta \theta$ and to the vibrational movement. In the approximation of a rigid C_∞ frame, six rotational and translational degrees of freedom of the XY are replaced by stretching and bending external vibrational modes. In real systems, the C_∞ lattice is not rigid so that the external modes of the molecule are coupled to vibrations of the cavity C_n coupled in turn to delocalized vibrations (phonons) of the infinite crystal lattice.

The naive pictures represented here must be replaced by the quantum mechanical treatment of such processes as the molecule–atom collision and photodissociation (predissociation) of molecule–atom complex. We will briefly resume one recently developed technique of calculations: the coupled-channel method [2].

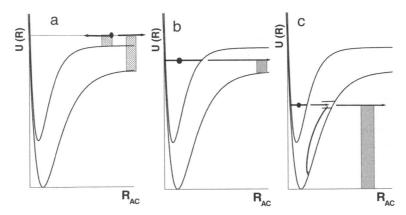

Fig. 4.2. (a) Collisional relaxation in the gas phase, (b) predissociation of a complex and (c) relaxation of the active center in a solid with dissipation of energy in the heat bath

The probabilities of transitions between the sets of initial $|i\rangle$ and final $|f\rangle$ states of the system are evaluated for a wave function with fixed angular momentum J,

$$\Psi^J = \sum_i^N \phi_i(R)g_i, \qquad (4.5)$$

where $\phi_i^J(R)$ is a translational function of the atom-to-molecule distance and g_i basis functions of the molecule constructed from its vibrational and rotational functions. The substitution of Eq. 4.5 into the Schrödinger equation yields a set of coupled equations,

$$d^2\phi_i/dR^2 = \sum_{i'} C_{ii'}\phi_{i'}(R) \qquad (4.6)$$

The $C_{ii'}$ matrix elements contain the coupling terms $\langle g_i|V|g_{i'}\rangle$, where $V(R, \theta, \phi, \ldots)$ is the intermolecular potential involving the polar angles supposed to be constant during the collision short as compared with the rotation period (the sudden approximation). The numerical solution of the coupled-equation system yields the rates of different $i \to f$ transitions, which must be averaged over the θ and ϕ polar angles and collision energies to be compared with the experimental data.

4.1.2 Experimental Studies of Relaxation Processes

General Remarks

Main Types of Experiments

We will consider three groups of relaxation processes:

1. The two-body $XY^{\#} + C \rightarrow (XY.C)^{\#} \rightarrow XY + C$ collisions (the *free → free* transitions): the vibrationally excited molecule $XY^{\#}$ forms a short-lived *collisional complex* $(XY.C)^{\#}$ with energy exceeding the XY.C bonding energy. Its dissociation yields the free XY molecule in its ground vibrational states, the energy $E_{XY^{\#}} - E_{XY}$ being redistributed between rotation and translation of XY + C. (Fig. 4.2 (a)).

2. The vibrational predissociation of the XY.C complex excited to its bound state $XY^{\#}.C$ with the vibrational energy of $XY^{\#}$ exceeding the bonding energy D_0 of XY.C in its ground state. The complex dissociates with the energy excess $E_{XY^{\#}} - D_0$ transferred to XY and C fragments (Fig. 4.2 (b)). This process (the bound → free $XY^{\#}C \rightarrow XY + C$ transition) is frequently called half-collision.

3. The bound → bound vibrational relaxation of an excited $XY^{\#}C$ complex or $XY^{\#}C_n$ cluster with energy below the dissociation threshold. The energy of $XY^{\#}$ is redistributed among the external modes of the cluster and increases its vibrational temperature. The same process occurs in an active center in the solid, which may be described as an $XY^{\#}C_n$ cluster within an infinite C_{∞} heat bath. The first relaxation step, $XY^{\#}C_n \rightarrow (XY.C_n)^{\#}$, is followed by a dissipation of the center energy in the C_{∞} heat bath (Fig. 4.2 (c)).

Real-Time Experiments

The real-time experiments involve a selective short-pulse excitation of a single (vibronic or ro-vibronic) level $|i\rangle$ of the molecule XY at $t = t_{pump}$ and probing the populations of the i level and of a set of final levels, $|f\rangle$, populated by the $i \rightarrow f$ relaxation as a function of the time delay, $\Delta t = t_{probe} - t_{pump}$. The observables are

- The decay rate of the i-level population $N_i(t)$ usually fitted by an exponential function,

$$N_i(t) = N_i(0)e^{-k_i t}. \tag{4.7}$$

- The population distribution among the f levels expressed by the branching ratio, P_f,

$$P_f = N_f(t_{probe}) / \sum_f N_f(t_{probe}). \tag{4.8}$$

The observed k_i decay rate is a sum of the intrinsic (radiative + nonradiative) decay rates of the i-state k_i^0 supposed to be environment-independent and of the relaxation (quenching) rate induced by the external perturbation k_i^q. The scope of the experiment is to determine $k_i^q = k_i - k_i^0$ and its dependence on the properties of the medium. k_i^q is the integral decay rate—the sum of differential state-to-state rates, which are not directly measured but are deduced from the observables,

$$k_{i \rightarrow f} = P_f k_i^q. \tag{4.9}$$

If all final levels are populated directly by an $i \rightarrow f$ transition, the time dependence of the N_f populations is given by

$$N_f(t) = N_i(0)(k_i^q / k_i) P_f (1 - \exp[-k_i t]). \tag{4.10}$$

The kinetics are more complex when relaxation is a sequential process involving intermediate states $|j\rangle$: $i \to j \to f$. The populations of intermediate states show an initial growth and a further decay, whereas population growth of final states is nonexponential and slower than the i-state decay.

XY + C Collisions in the Gas Phase

The XY-C interaction is significant only for the XY-C distances where R_{XY-C} does not exceed the interaction limit R_{int}. In the gas phase, the number of XY+C collisions per second, n_{coll} s^{-1}, may be roughly approximated as

$$n_{coll} = k_{coll} N_C \approx \sigma_{XY-C} \langle v_{XY-C} \rangle N_C, \qquad (4.11)$$

where N_C is the number of molecules of C per cm^3, $\sigma_{XY-C} = 2\pi R_{int}^2$ is the cross-section of XY + C collisions and $\langle v_{XY-C} \rangle$ is the average velocity of the XY + C pair. On the other hand, if XY$^\#$ and C do not form long-lived complexes, the duration of a collision is of the order of the transit time through the interaction region: $\tau_{coll} \approx 2R_{int}/\langle v_{XY,C} \rangle$.

At $T = 300$ K, $\langle v_{XY,C} \rangle \approx 3 \times 10^4$ cm/s for molecular masses $M_{XY} \approx M_C \approx 50$ so that $\tau_{coll} \approx 3 \times 10^{-12}$ s whereas the average between-collision time, $t_{coll} = 1/n_{coll}$, is $\sim 10^{-7}$ s at the gas pressure of 10^{-3} bar. The ratio of molecules involved simultaneously into an XY+C collision is thus of the order of $\tau_{coll}/t_{coll} \approx 10^{-5}$, and the direct observation of XY$^\#$C collisional complexes is extremely difficult. The collision effects are deduced from population transfer due to collisions occurring at $\Delta t = t_{probe} - t_{pump}$ observation time. Δt and N_C are chosen in a way as to avoid multiple collisions.

The efficiency of a relaxation process is currently expressed in terms of the ratio of the observed k_i^q quenching rate and the elastic collision rate in the hard-sphere approximation k_{coll}^{HS}. An equivalent procedure is the expression of the collision efficiency by its *effective cross-section*, $\sigma^q = k^q/\langle v_{XY-C} \rangle$. Both procedures are convenient because they allow separating the efficiency of collisions from their number, which depends on the $\langle v_{XY-C} \rangle = \sqrt{8kT/\pi \mu_{XY-C}}$, i.e., on the masses of the collision partners and on the temperature of the gaseous mixture. They must, however, be used with caution because they are based on the assumption that the efficiency of a collision (cross-section) is independent from the collision energy, i.e., of the velocity, v_{XY-C}. The cross-section is energy independent for purely repulsive hard-sphere potential but varies with energy when the attractive interactions are taken into account in Lennard-Jones and more sophisticated potentials [1].

For a detailed discussion of collisional mechanisms, see [3, 4].

XY.C Complexes, XY.C$_n$ Clusters and XY.C$_\infty$ Solids

The common feature of these systems is the coupling between internal vibrational modes of XY with external vibrational modes of the complex (cluster). The initially excited bright $|s\rangle$ $|v_{XY}\rangle$ level is coupled to a set of $|\ell\rangle$ levels: $|v'_{XY} v_\sigma v_\pi\rangle$ levels, where $v'_{XY} < v_{XY}$ whereas v_σ and v_π indicate excitation of external σ and π modes. Some of

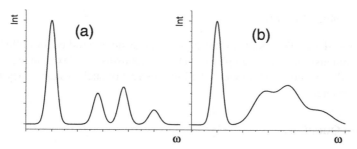

Fig. 4.3. The satellite bands of the $0 \rightarrow s$ transition in the spectrum of (a) a free complex and (b) the active center in the rare-gas matrix

these levels are not completely dark and give rise to the satellite bands accompanying vibronic transitions $v_{XY} = 0 \rightarrow v$ in absorption (excitation) spectra of complexes or solids (Fig. 4.3).

If the energy $E_s \approx E_\ell$ exceeds the threshold of dissociation of an isolated XY.C complex, the s- and ℓ-levels are coupled to the dissociative continuum $\{m\}$. In an active center in the solid, all excited levels are coupled to the heat bath—the quasi-continuum of phonon states of the C_∞ crystal lattice. The information about the strength of the $s \leftrightarrow \{m\}$ and $\ell \leftrightarrow \{m\}$ coupling strength may be deduced from the homogeneous widths of $0 \rightarrow s$ and $0 \rightarrow \ell$ lines in the absorption (or fluorescence-excitation) spectra of a complex (or solid). In the spectra of free complexes, the widths of $0 \rightarrow s$ and $0 \rightarrow \ell$ transitions are not very different (Fig. 4.3 (a)), whereas the absorption spectra of XY-C_∞ mixed crystals are composed of narrow zero-phonon lines ($0 \rightarrow s$ transitions) and broad and often structureless phonon bands corresponding to overlapping broad $0 \rightarrow \ell$ lines (Fig. 4.3 (b)). The increased widths of $0 \rightarrow \ell$ lines indicate that the $|\ell\rangle$ levels are more strongly coupled to $\{m\}$ than the $|s\rangle$ levels involving only the internal modes.

The studies of XY.C complexes and XY.C_n solids are complementary. The time evolution of an XY$^\#$C complex is interrupted by its dissociation resulting from the transfer of internal energy of excited XY molecules to external modes. The energy distribution between vibrational, rotational and recoil energy of the XY + C disso-ciation products contains information about the predissociation mechanism. In the active center in a solid, this energy is rapidly evacuated to the heat bath, so that the initial structure and temperature of the XY$^\#$C$_n$ center is recovered, which allows the study of the further steps of the cascade of $v \rightarrow v - 1 \rightarrow v - 2 \cdots$ processes in the unchanged environment.

The case of XY.C_n finite clusters is more complex. The $E_v - E_{v-1}$ energy is not evacuated, so that the vibrational temperature of the cluster is increased. This may induce a change of the cluster structure (melting). The weakly bound cluster of rare-gas atoms loses slowly its energy excess by evaporation of one or several C-atoms, $XYC_n \rightarrow XYC_{n-1} + C$. The dynamics of such systems is complex and it is practically impossible to separate the vibrational relaxation of the solute from the evolution of the cluster structure.

Hydrogen-Bonded Systems

We will extend the discussion to more complex systems studied in collision-free conditions but also in condensed phases. The mechanisms of relaxation are the same as in the model systems and, in view of a high interest of such systems, they must be briefly treated here.

4.1.3 Main Questions

We will focus the further discussion on three questions that seem to be the keys for a better understanding of relaxation mechanisms:

1. Is it possible to apply, as the first approximation in the study of relaxation processes, the picture of weakly communicating degrees of freedom (rotation, normal vibrational modes) widely used in spectroscopy?
2. How important are sequential mechanisms involving external degrees of freedom (external modes) in the vibrational predissociation and relaxation?
3. Is it possible to establish the propensity rules for competing relaxation channels independent of the specific properties of molecules?

The experimental data for $XY^{\#} + C$ collisions, $XY^{\#}C$ half-collisions and $XY^{\#}C_{\infty}$ centers in rare-gas matrices will be limited to a few model systems, the light diatomic (hydrides) and small polyatomic molecules and a few larger ones of a direct chemical interest. On the other hand, we are interested in the systems with low vibrational energies, i.e., with discrete vibrational-level sets.

4.2 Collisional Vibrational and Rotational Relaxation

The decay of the initially excited ro-vibronic state vJ is described by the rate constant

$$k_{vJ}^{q} = k_{vJ}^{rot} + k_{vJ}^{vib} = \sum_{J'} k_{vJ-vJ'} + \sum_{v'} \sum_{J'} k_{vJ-v'J'} \qquad (4.12)$$

which is the sum of differential rate constants $k_{vJ \to vJ'}$ of the purely rotational relaxation and $k_{vJ \to v'J'}$ for the vibrational relaxation ($v' \neq v$). Upon a nonselective excitation of all rotational levels of a v-state, only the J-averaged constant of vibrational relaxation,

$$k_{v}^{vib} = \langle k_{vJ}^{vib} \rangle_{J}, \qquad (4.13)$$

may be measured.

Two other relaxation processes must be taken into account:

- When the initial velocity distribution is not equilibrated (as in the case of a narrow-band excitation of a limited class of velocities [5]), the equilibration of velocities is established with the k^{trans} rate. This process may be monitored by measurements of the Doppler widths of the absorption or emission lines.

- The optical excitation with the natural or polarized light prepares a partial alignment of molecular axes. The alignment (which may be probed by polarization measurements) is reduced by the collision-induced reorientation of molecular axes (the Δm_J transitions) with the k^{or} rate. The probability of reorientation of the rotation axis is, however, small as compared with collision-induced change of the angular velocity as it is a rule for all rotating tops [4].

Finally, $k^{trans} \gg k^{rot}$ and $k^{or} \ll k^{rot}$, so that these processes do not strongly interfere with the measurements of rotational relaxation.

4.2.1 Rotational and Vibrational Relaxation Rates

A very large amount of experimental data is actually available. Some of them are discussed in the review papers centered either on the diatomic [6–8] or on polyatomic [9–10] molecules.

There is a big difference between rotational and vibrational relaxation rates. The rate constants k^{rot} (cross-sections σ^{rot}) of rotational relaxation are nearly the same in a large class of XY+C pairs and close to the gas-kinetic collision rates whereas the vibrational relaxation rates are smaller by orders of magnitude and drastically different for different XY+C systems. For instance, for OH $X^2\Pi$ ground state at 300 K $k^{rot} \approx 10^{-10}$ for Ar, N_2 and O_2 and $\sim 4 \times 10^{-10}$ s^{-1} molecule^{-1} cm^3 for H_2O [11] close to the Lennard–Jones collision rate of $k^{LJ}_{coll} \approx 5 \times 10^{-10}$ s^{-1} molecule^{-1} cm^3. In the same conditions, the vibrational relaxation rates of HF vary from $k^{vib} = 6 \times 10^{-15}$ to 1.2×10^{-12} s^{-1} molecule^{-1}cm^3 for He and CO_2 as collision partners [12, 13]. The ratio $k^{vib}/k^{rot} \leq 10^{-4}$ typical for light diatomic molecules is increased to $10^{-2} - 10^{-1}$ for small polyatomics. For instance, $k^{rot} = 3.45 \times 10^{-10}$ and $k^{vib} = 3 \times 10^{-9}$ s^{-1} molecule^{-1} cm^3 for the 8^1 level of glyoxal CHO-CHO in the S_1 excited state with CO as the collision partner [14].

The $k^{vib}/k^{rot} \ll 1$ ratio cannot be explained uniquely by different energy gaps between the neighbor vibrational and rotational levels because it is maintained for high overtone levels of light polyatomics (HCN, acetylene), where the spacing between rotational and vibrational levels is nearly the same [15-17]. The probability of transitions between closely spaced vibrational levels is reduced because they correspond to combinations of different normal modes so that such transitions involve large changes of vibrational quantum numbers (see Sect. 4.4.1).

A good agreement between the rate constants or cross-sections for rotational relaxation and for elastic collisions indicates that rotational transitions necessitate a close approach of A and C molecules. A large enhancement of k^{rot} is observed for collisions between two strongly polar molecules. So, k^{rot} values vary with J and K between 2.7 and 3.5×10^{-9} in the 2_44_4 state ($E_{vib} \approx 11300$ cm^{-1}) of D_2CO [18] and between 3.7 and 6.3×10^{-9} s^{-1} molecule^{-1} cm^3 in the 3_3 state of HCN ($E_{vib} = 9622$ cm^{-1}) [18]. A significantly lower value of $k^{rot} = 7.5 \times 10^{-10}$ s^{-1} molecule^{-1} cm^3 was reported for $C_2H_2^{\#} + C_2H_2$ collisions in the 3_3 state of acetylene ($E_{vib} = 9637$ cm^{-1}). This difference suggests a predominant role of long-range electric dipole–dipole interactions in rotational relaxation of polar species. This

Table 4.1. k_0 (in $\mathrm{torr^{-1}s^{-1}}$ units) and α parameters for rotational relaxation of acetylene

Level	E (cm^{-1})	Transition	k_0	α	Ref.
3_1	3287	$J \to J'$	8.1	1.92	15
3_1	3287	$II, J \to I, J'$	1.3	1.7	
3_3	9637	$J \to J'$	6.4	1.68	16
3_3	9637	$v, J \to v', J'$	0.17	1.05	
$2_1 3_3$	11600	$J \to J'$ (fast)	9	1.6	20
$2_1 3_3$	11600	$J \to J'$ (slow)	4	1.6	
$2_1 3_3$	11600	$v, J \to v', J'$ (fast)	0.6	1.1	
$2_1 3_3$	11600	$v, J \to v'J'$ (slow)	0.45	1.1	

suggestion is corroborated by a propensity for small ΔJ changes, which seems to be due to a memory of the $\Delta J = \pm 1$ selection rule for radiative electric dipole transitions.

4.2.2 Propensity Rules for Rotational Transitions

The dependence of the rates of $J \to J'$ collision-induced transitions in the room-temperature gas on the momentum (ΔJ) and energy (ΔE) gaps between initial and final levels is nearly the same for diatomic (CO [19], NH [20], etc.) and small poly-atomic molecules (acetylene [15-16], glyoxal [14, 21]). The limits of ΔJ and ΔE changes are wide: for instance, ΔJ up to 20 and $|\Delta E|$ up to 600 cm$^{-1} \approx 3kT$ are reported for high vibrational levels of the ground-state acetylene molecules [14–16]. The rates vary slowly with the initial J value but strongly depend on ΔJ and ΔE. Because ΔJ is limited by conservation of the total kinetic moment $\mathbf{J}_{XY} + \mathbf{J}_{XY\text{-}C} + \mathbf{J}_C$, a correlation between $k_{J \to J'}$ and ΔJ could be expected but it was reported only for collisions of two strongly polar molecules [18]. Otherwise, the values of $k_{J \to J'}$ differential rate constants correlate much better with ΔE than with ΔJ. This correlation is relatively well described by the EGL (exponential gap law) relation between the rate and the absolute value of the energy gap:

$$k_{JJ'} = k_0 e^{-\alpha|\Delta E|/\langle E_{\text{coll}}\rangle} = k_0 e^{-\alpha|\Delta E|/kT} \tag{4.14}$$

in the case of collisions of acetylene [10], CO [19], NH [20] and glyoxal ($S_1(A_u)$) [21] with different collision partners. The k_0 and α parameters for purely rotational transitions and for transitions between rotational levels of different vibrational states $v \to v'$ or different components of Fermi doublets ($I \to II$) are listed in Table 4.1 for several excited levels of acetylene.

The agreement with experiments may be improved by addition of the power factor (the PEGL formula):

$$k_{JJ'} = k_0(|\Delta E_{JJ'}|/B)^b e^{-\alpha|\Delta E_{JJ'}|/kT}, \tag{4.15}$$

but the physical meaning of this purely empirical formula is not evident. It is important

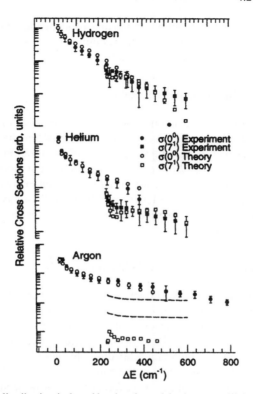

Fig. 4.4. Population distribution induced by the glyoxal-hydrogen collisions in a crossed-beam experiment (from [21])

to note that, in Eqs. 4.14 and 4.15, the $2J + 1$ spatial degeneracy of rotational states is not taken into account. The ratios of the observed rates of the $J \to J'$ and reverse $J' \to J$ transitions rates show that $k(J \to J')/k(J' \to J)$ is close to $e^{-\Delta E/kT}$ without the $(2J+1)/(2J'+1)$ factors. This result indicates that the spatial orientation of the **J** vector is conserved in collisions, which change strongly its length $|\mathbf{J}|$ and is consistent with the previously mentioned $k^{\mathrm{rot}} \gg k^{\mathrm{or}}$ relation.

The data of Table 4.1 show that the propensity rules (expressed by the α factors) are nearly the same for $vJ \to v'J'$ transitions between different vibrational states as for the purely rotational $vJ \to vJ'$ relaxation. The rotation–translation coupling seems to be independent of the vibration–translation interaction. The same propensity was observed for glyoxal excited to $K = 0$ levels of $S_1 0^0$ or 7^2 vibronic state and submitted to a single collision with H_2 or the rare-gas atoms in a crossed-beam experiment [21]. The populations of K-levels resulting from the purely rotational $(0^0, K = 0 \to 0^0, K)$ and from ro-vibrational transitions $(0^0, K = 0 \to 7^1, K)$ or $(7^2, K = 0 \to 7^1, K)$ are the same and may be fitted by the EGL formula (Fig. 4.4).

Such a behavior suggests a picture of two quasi-independent processes involving rotation–translation and vibration–translation coupling taking place during a collision without a significant vibration–rotation coupling.

4.2.3 Vibrational Relaxation and Up-Relaxation

We will try to extrapolate the last conclusion to the problem of the vibrational transitions and to check whether the energy exchange between the vibrations of the molecule and its heat bath is determined by its overall vibrational energy content: $E_{vib} = \sum_i E_i$ or by the energy E_i of each i-mode. In the first model, one can expect that the hot molecule will lose a fraction of its vibrational energy in each collision with a cold collision partner. In the latter one, the overall energy is not important: the system tends to establish the thermal equilibrium between the populations of the $v_i = 0, 1, \ldots$ levels of each mode.

Vibrational Excitation in Gas-Phase Collisions

In the early pioneering works, the effects of collisions were investigated for several medium-sized molecules (benzene [22], aniline [23], p-difluorobenzene (p-DFB) [24]) excited to higher vibrational levels of the S_1 state in the room-temperature gas. Their surprising result was that instead of the vibrational cooling, an efficient up-relaxation was observed. So, the main relaxation channel of benzene excited to the $6^1 (E_{vib} = 522 \text{ cm}^{-1})$ vibronic state is the endoergic $6^1 \to 6^1 16^1$ transition with $\Delta E = +237 \text{ cm}^{-1}$. Similar relations were found for other molecules.

The branching ratios for transitions from different levels of the S_1 state of p-DFB are given in Table 4.2. The $\Delta v_{30} = +1$ up-relaxation involving the lowest frequency $\omega_{30}= 119 \text{ cm}^{-1}$ out-of-plane mode is the main channel for a whole set of vibrational levels independent of the excitation of other modes that behave as spectators. The ratio of $\Delta v = +1/\Delta v = -1$ transition probabilities from the 30^1 and 30^2 states is very significant. It indicates that the system tends to establish a thermal equilibrium between populations of the 30^v levels, which corresponds at $T = 300 \text{ K}$ to $N(v_{30}) = 1/N(v_{30} = 0) \approx N(v_{30} = 2)/N(v_{30} = 1) \approx 0.56$.

These data suggest the coupling between each individual vibrational mode and the continuum of translational energies. The absence of transitions between close-lying DFB levels involving different modes, as e.g., the $8^1 \to 30^1$ transition with $|\Delta E|= 54 \text{ cm}^{-1}$ shows that collision-induced coupling between different vibrational modes is weak. To our best knowledge, there is no indication either of an efficient, collision-induced vibration-to rotation energy transfer.

The collisional up-relaxation is highly selective and limited to excitation of one or a few low-frequency modes. In the crossed-beam experiments, collisions between p-DFB and He or Ar atoms with the energies between 150 and 1,500 cm^{-1} exceeding the excitation onset of a number of modes induce only the $\Delta v_{30} = +1$ and +2 transitions, whereas the 8^1 level with energy intermediate between 30^1 and 30^2 is not excited [25]. Only the $0^0 \to 7^1$ and $7^2 \to 7^1, 7^3$ transitions involving the lowest

Table 4.2. Branching ratios for collision-induced transitions from several levels of p-DFB + He ($T = 300$ K)

Level	E (cm^{-1})	$\Delta v_{30} = +1$	$\Delta v_{30} = -1$	$\Delta v_8 = +1$	$\Delta v_8 = -1$	Other
0^0	0	0.71	—	0.23	—	0.06
30^1	119	0.50	0.28	0.09	0.13	
8^1	173	0.36	—	0.12	0.28	0.24
30^2	238	0.26	0.48	0.08	—	0.18
$30^1 8^1$	292	0.20	0.36	0.06	0.15	0.23
6^1	410	0.56	—	0.10	—	0.34
5^1	818	0.65	—	0.08	—	0.26

frequency mode $\omega_7 = 230$ cm^{-1} are observed upon the crossed-beam collisional excitation of glyoxal in the $S_1^1(A_u)0^0$ state [21].

The strong coupling of the ω_{30} DFB and ω_7 glyoxal modes is not due to their out-of-plane character, as it was initially supposed. The coupled channel calculations of DFB up-relaxation [2] show that, in a virtual DFB molecule, in which the frequency of one of ω_X, $X = 6, 8, 17, 22$ and 27 modes, is put equal to ω_{30}, the probability of the $0^0 \rightarrow X^1$ transition is the highest one and nearly independent of the X-mode symmetry. As previously mentioned (Sect. 4.1), the strength of the vibration-translation coupling is related to the frequency ratio between an internal and an external mode of the collisional complex and is the strongest for the low-frequency internal modes, in agreement with the energy gap law.

The Energy Transfer from the Vibrational Continuum

The vibrational and rotational excitation of small molecules (energy acceptors) (CO_2, H_2O) induced by collisions with vibrationally hot ($E_{vib} = 36,000$ to $42,000$ cm^{-1}) but rotationally and translationally cold medium-sized molecules (energy donors) such as pyrazine [26–28] is a specific case of the V-V energy transfer: the energy transferred to internal degrees of freedom of acceptors is not provided by the continuum of translational energies but by a quasi-continuum of vibrational energy sink of the donor.

The energy distribution in CO_2 excited by collisions with hot pyrazine is given in Table 4.3, where J_{max} indicates the rotational level with maximum final population, k_{xJ} are the rate constants of transitions to individual rotational levels and k_{xv}, the integrated transitions rate to the vibrational state, v. In view of a large thermal population of the low J-levels of the $(00^0 0)$ state before collision, the J_{max} for this state cannot be determined.

These data suggest the existence of two distinct $V \rightarrow R, T$ and $V \rightarrow V, R, T$ channels:

- High rotational levels of the vibrationless state of CO_2 are populated with an overall rate of the order of the collision rate. The rotational temperatures (estimated

Table 4.3. Nascent population of CO_2 levels induced by pyrazine ($E_{vib} = 40,600$ cm^{-1}) at $T = 243$ K

Level	J	k_{xJ}^*	k_{xv}^*	T_{rot} (K)	T_{trans} (K)
(00^00)	60	$6.5\ 10^{-12}$			1620
	70	$4.0\ 10^{-12}$	$\sim 10^{-10\dagger}$	1200	3100
	80	$2.1\ 10^{-12}$			5500
	J_{max}	$k_{xJ_{max}}$			
(00^01)	15	$4.6\ 10^{-13}$	$5.6\ 10^{-12}$	318	380
$(10^00/02^00)^\ddagger$	22	$5.8\ 10^{-13}$	$8.4\ 10^{-12}$	315	415
$(10^00/02^00)^\ddagger$	18	$5.0\ 10^{-13}$	$4.8\ 10^{-12}$	390	395
(02^20)	10	$6.5\ 10^{-13}$	$1.6\ 10^{-11}$	296	380

*In s^{-1} molecule^{-1} cm^3 units.
†Tentative estimation by the authors of this review.
‡Components of the doublet resulting from the $(10^00) + (02^00)$ Fermi resonance.

from the population distribution in $J > 65$ levels) and translational temperatures (deduced from linewidths) are very high.

- The $V \rightarrow V$ transfer populating the (00^01), $(10^00)/02^00)$ and (02^20) levels of CO_2 is less efficient by an order of magnitude and not accompanied by a significant increase of rotational and translational energy of acceptor.

The amount of experimental data is still limited, but they seem to open a new insight into the vibrational and rotational excitation mechanisms.

4.2.4 Conclusion

The experimental data are consistent in the first approximation with a simple model of noncommunicating internal degrees of freedom of the molecule: rotation and a set of vibrational modes. Each of them is coupled to the continuum of translational energies (the heat bath), this coupling being stronger than the coupling between molecular degrees of freedom. The rotation–translation coupling is the strongest one, as evidenced by a high efficiency of collisional rotational relaxation. The vibration–translation coupling is weaker and its strength decreases with increasing frequency of the vibrational mode. In a supersonic jet, such molecules with high vibrational frequencies as CO are rotationally cold but the thermal populations of $v \geq 1$ levels remain practically unchanged.

The rotation–vibration coupling and the coupling between vibrational modes is much weaker. To our best knowledge, the loss of a large amount of the vibrational energy in an $XY^\# + C$ collision is not accompanied by a significant increase of its rotational energy, i.e., by the vibration-to-rotation transfer. There is either no indication of an efficient collision-induced energy transfer between the normal modes.

4.3 Direct and Sequential Paths of the Vibrational Relaxation

4.3.1 Model Notions

There is a close analogy between the vibrational relaxation of the excited $XY^{\#}$ molecule induced by an $XY^{\#}+C$ collision, by predissociation of an isolated $XY^{\#}C$ complex and by dissipation of the energy of an $XY^{\#}C_n$ center in its heat bath: the infinite C_{∞} crystal. The case of $XY.C$ complexes and $XY.C_n$ active centers is better defined than that of collisional interactions. The initial configuration and bonding energy of the system in its ground state is known. The strength of the $XY^{\#}$-C interaction energy in the excited state and the frequencies of external modes σ and π may be deduced from absorption spectra so that the essential information about the excited level pattern is available.

In the C_{∞} solid, all $E_{\text{vib}} > 0$ levels of the $XY.C_n$ center are instable in view of their coupling to the quasi-continuum $\{m\}$ of phonons. The dissociative continuum of energies of $XY + C$ fragments plays the same role in the case of excited levels of a free $XY.C$ complex with the energy exceeding its dissociation threshold ($E_{\text{vib}} > D_0$). The vibrational predissociation of collision-free complexes and vibrational relaxation in rare-gas matrices will be thus treated in the same way: in terms of the coupling between initially excited bright $|s\rangle$ level, discrete dark $|\ell\rangle$ levels and the $\{m\}$ continuum or quasi-continuum.

The model system is a diatomic molecule XY forming a complex with a rare-gas atom C, with a diatomic molecule CD or embedded in a rare-gas matrix C_{∞}. We will limit our discussion to two types of molecules: the hydrides HF and OH with high ($\omega_{XY} \geq 3,600 \text{ cm}^{-1}$) stretching frequencies (the experimental data are given in Tables 4.4 and 4.5) and the rare-gas complexes of the electronically excited iodine molecule $I_2^* - {}^3B_{0u^+}$, where $\omega_{XY} = 126 \text{ cm}^{-1}$. We will then extend it to more complex systems: the van der Waals and hydrogen-bonded complexes of medium-sized polyatomic molecules in order to show the validity of the rules deduced from the study of diatomics.

4.3.2 The Vibrational Relaxation of Diatomics

Atom–Diatom Complexes

The simplest case is that of a triatomic $XY.C$ complex (where C is a rare-gas atom), the vibrational spectrum of which is composed of an internal X-Y stretching mode and of two external modes: the σ $XY..C$ stretching and π bending mode. The frequency of the internal mode ω_{XY} is significantly higher than the $XY..C$ dissociation threshold of Ar..HF and of I_2^*.He complexes, so that the $|s\rangle - v_{XY} = 1$ level is directly coupled to the continuous part of the σ mode energy spectrum. On the other hand, the excited levels of the π mode with energies close to that of the v_{XY} form a set of discrete $|\ell\rangle - v_{\pi}$ levels. Their energies are much higher than the energy barrier for internal rotation of X-Y in the X-Y-C plane, so that they may be approximated by those of a free XY rotator: $E(v_{\pi}) \approx B_{XY} J(J + 1)$ with $J = v_{\pi}$. Because $E(v_{\pi}) > D_0$, the v_{π} levels are *predissociated by rotation* [29].

Two relaxation channels are thus open:

- a *direct vibrational predissociation* path resulting from the coupling of the v_{XY} state to the XY+C continuum and
- a *sequential process* $s \to \ell \to \{m\}$, in which the transition from the v_{XY} level to the resonant $v_{\pi} \gg 1$ level of the external bending mode is followed by dissociation. When $E(v_{\pi}) \gg D_0$, a fraction of this energy induces dissociation of the complex, the remaining part being shared among the rotation of XY and recoil energy of XY and C fragments. One can represent this process as a series of $v_{\pi} \to v_{\pi} - 1$ steps in which the energy is transferred from the bending to the XY..C stretching until the dissociation limit is attained. The major part of the $E(v_{\pi}) - D_0$ energy is conserved in XY rotation.

One can presume that the efficiency of the relaxation channels is governed by momentum (energy) gap laws. The rate of the direct predissociation is expected to decrease with increasing amount of energy transferred to translation $\Delta E = E(v_{XY}) - D_0$ and with increasing frequency ratio $R = \omega_{XY}/\omega_{\sigma}$. One can also suppose that the $\ell \to \{m\}$ step of the sequential process (dissociation of $v_{\pi} \gg 1$ levels) is rapid in view of small Δv and ΔE changes in each $v_{\pi} \to v_{\pi} - 1$ step. The rate of the sequential process would be thus limited by the rate of the intracomplex vibrational $v_{XY} \to v_{\pi}$ redistribution which depends on the value of the $\Delta v_{tot} = v_{XY} + v_{\pi}$ parameter.

The dissociation rates of HX.C complexes are extremely low: only the lower limit of the decay time of Ar.HF ($v = 1$), Kr.HF ($v = 1$) and Ar.OH ($v = 2$) could be determined [30–32] (Table 4.4). Their lifetimes are of the order of 10^{12} classical periods of HX vibrations. The decay times of excited vibrational levels of $I_2 B\,^3\Pi_{0u+}$ He, Ne and Ar complexes deduced from their line widths are relatively short, of the order of 10^{-10}s [33], but they correspond still to more than 10^3 vibrational periods of I_2.

The striking difference between hydride and iodine complexes may be explained by assuming that the direct mechanism is predominant, as shown in the pioneering theoretical studies of iodine complexes [34], which have been the starting point of the treatment of vibrational relaxation in terms the energy gap law. The direct mechanism of I_2.C predissociation was confirmed for the linear isomer of I_2.Ar by the energy distribution in the dissociation products with its small fraction transferred to the rotation of I_2 [35]. The predominant role of the direct predissociation channel is consistent with the relatively low value of the frequency ratio $R \approx 10$–15, whereas the s-ℓ channel seems to be extremely inefficient: the rotational constant of heavy iodine molecules ($B = 0.0374$ cm^{-1} is so small that the v_b bending level resonant with $v_{XY} = 1$ corresponds to $J \approx 250$!

In contrast with I_2.C, the direct coupling between internal and external stretching in hydrides must be extremely weak in view of large values of R: e.g., in the Ar..HF complex $R \approx 80$ [30]. The direct relaxation channel is thus practically closed. It is, however, surprising that the sequential path involving the bending mode is so inefficient, whereas its efficiency is high in the case of HF and OH complexes with diatomic molecules.

Table 4.4. Lifetimes of ground-state OH and HF complexes

Compound	Mode	Level (v)	τ (sec)	Measure	Ref.
Ar.HO	H-O	2	$> 6 \times 10^{-6}$	D*	30
Ar.HF	H-F	1	$> 3 \times 10^{-4}$	D*	32
Kr.HF	H-F	1	$> 3 \times 10^{-4}$	D*	32
H_2.HO	H-O	2	1.15×10^{-7}	D	40
D_2.HO	H-O	2	$< 5\ 10^{-9}$	D	40
			$> 5 \times 10^{-11}$	L	42
CH_4.HO	H-O	1	3.8×10^{-11}	L	41
CH_4.HO	H-O	2	2.5×10^{-11}	L	41
H_2.HO	H-H	2	$< 7 \times 10^{-9}$	D	40
o-H_2.HF	H-F	1	2.7×10^{-8}	L	36
o-D_2.HF	H-F	1	1.12×10^{-9}	L	36
p-D_2.HF	H-F	1	1.4×10^{-9}	L	36
D_2.DF	D-F	1	3.5×10^{-8}	L	36
NO.HF	H-F	1	0.8×10^{-9}	L	38
CO_2.HF	H-F	1	2.7×10^{-8}	L	37

D, from measurements of dissociation rates.
L, from linewidth measurements.
*Only the lower-limit estimation.

XY.CD Complexes and XY.C_n Centers in Solids

We have at our disposal a large amount of data on vibrational relaxation of hydrides in jet-cooled complexes with diatomics and in rare-gas matrices. Their relaxation is rapid compared with that of rare-atom complexes; this acceleration may be tentatively explained by an increased number of external modes. The sequential relaxation mechanism is predominant and the difference of rates may be qualitatively explained by differences in Δv_{tot} parameters in agreement with the momentum gap law.

Δv_{tot} Dependence of Relaxation Rates

The intracomplex redistribution of energy, $v_{XY} \rightarrow v'_{XY} v_\sigma v_\pi$, is obviously the rate-determining step, and the favored relaxation channels correspond to the minimum values of the Δv_{tot} parameter. For this reason, the $\Delta v_{XY} = -1$ channel is always preferred to $\Delta v_{XY} = -2$. Upon the excitation of the $v_{\text{H-H}} = 2$ or $v_{\text{OH}} = 2$ levels of the H_2.HO complex, only the $v = 1$ level of the OH fragment and not the $v = 0$ one is populated [39]. If one assumes that the whole energy of the OH stretching (3,600 cm^{-1}) is transferred to internal rotation, then the rate of the $v_{\text{OH}} = 2 \rightarrow v_{\text{OH}} = 1 v_\pi = 14$ transition is limited by a large Δv_{tot} value. The long (115-ns) lifetime of the $v_{\text{OH}} = 2$ level is reduced in the D_2.HO complex in which $\omega_{D_2} < \omega_{\text{OH}}$, whereas $\omega_{H_2} > \omega_{\text{OH}}$ (Fig. 4.5 (left)). This reduction is due to a transfer of the major part of the OH stretching energy to the D_2 vibration. Only 600 instead of 3,600 cm^{-1} goes

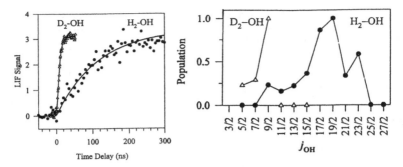

Fig. 4.5. The growth of the population of the $v = 1$ level of free OH fragments and rotational population distribution in OH formed by dissociation of OH.H$_2$ and OH.D$_2$ complexes excited to the $v_{OH} = 2$ level (from [40])

to external modes, which implies $\Delta v_{tot} = 7\text{–}8$ instead of 15 [40]. As shown in Fig. 4.5 (right), only the lowest rotational levels of the OH ($v = 1$) fragment issued from the OH.D$_2$ are populated whereas the energy content is much higher in OH formed by dissociation of the OH.H$_2$ complex. The same difference is observed in the case of HF.H$_2$, DF.D$_2$ and HF.D$_2$ complexes [36].

A further acceleration of relaxation processes is reported for the complexes of polyatomics: the lifetimes of $v_{OH} = 1$ and 2 levels of CH$_4$.HO are shorter than those of the D$_2$ complex [41] in view of a rapid redistribution of the energy of the OH stretch among the fundamentals of stretching and overtones of bending modes of CH$_4$ in the 2,500–3,000 cm^{-1} frequency range.

Stretch–Bend (Rotation) Coupling

The sequential relaxation mechanism seems to be predominant in all XH.CD and XH.C$_n$ systems. A direct proof of the efficiency of the vibration-to-internal rotation energy transfer is a large fraction of the energy excess $E_{vib} - D_0$ transferred to rotation of the XY and CD fragments. So, upon the excitation of the $v_{OH} = 2$ level of the H$_2$·OH complex, the rotational levels of the OH fragment up to $J = 23/2$ are populated with the average rotational energy $E_{rot} \approx 2,300$ cm^{-1}—about 2/3 of the overall energy, its remaining part being divided between H$_2$ rotation and recoil energy [40]. A similar energy distribution is reported for HF complexes with N$_2$, CO and NO [43–44]. For the first of them, the main relaxation channel,

$$N_2 \cdot HF\ (v_{HF} = 1) \rightarrow N_2\ (v = 0, J \leq 13) + HF\ (v = 0, J = 12),$$

corresponds to 90% of free energy in the HF rotation, 5% in N$_2$ rotation and 5% in translation.

XH Stretch–Rotation Coupling in Solids

The role of quasi-rotational levels as intermediates of the vibrational relaxation is also evidenced by the apparent breakdown of the energy gap law for vibrational relaxation of XH.C$_n$ centers in matrices.

Table 4.5. Energy gaps (cm^{-1}) and relaxation rates of HX in matrices

Guest	v	$\Delta E_{v,v-1}$	Host : Ne	Ar	Kr	Ref.
OH($A^2\Sigma^+$)	1	2980	0.9×10^5	$> 5 \times 10^7$		48
OD($A^2\Sigma^+$)	1	2268	0.39×10^5			
OH($A^2\Sigma^+$)	2	2790	4.0×10^5	$> 5 \times 10^5$		
OD($A^2\Sigma^+$)	2	2112	1.4×10^5			
HCl($X^1\Sigma^+$)	1	2885		8.3×10^2	1.3×10^3	49
HCl($X1\Sigma^+$)	2	2781		4.1×10^3	1.0×10^4	
HCl($X^1\Sigma^+$)	3	2637 6.2×10^3	5.6×10^4	3.4×10^5		
DCl($X^1\Sigma^+$)	2	1980		1.2×10^2		

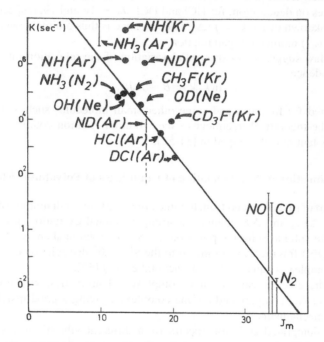

Fig. 4.6. The vibrational relaxation rates of small molecules in rare gas matrices plotted against the J_m parameter

The extremely small ($\sim$1 s^{-1}) nonradiative relaxation rates of $v = 1$ levels of light molecules, such as N$_2$ or CO [45, 46] up to $k_{\rm rel} \approx 4 \times 10^{-3}$ s^{-1} for O$_2$ [47], may be explained by assuming the direct coupling between high-frequency internal modes ($\omega \approx 1{,}600$–2,300 cm^{-1}) and $\omega \le 50$ cm^{-1} lattice (phonon) modes corresponding to $\Delta v_{\rm tot}$ of the order of 40. The direct relaxation model cannot, however, explain the large decay rates of HX hydrides with energy gaps between vibrational levels $\omega_{\rm HX} \ge 2{,}800$ cm^{-1} much larger than those of N$_2$ or CO. Still more surprising are the rates of deuterated DX species reduced as compared with HX in spite of their lower frequencies ($\omega_{XH}/\omega_{XD} \approx \sqrt{2}$) [48–50]. This difference is consistent with the assumption of the sequential relaxation mechanism involving, as the rate-determining step, the energy transfer from the v_{XH} level to close-lying $v_\pi \gg 1$ levels of the bending mode approximated by those of the free rotor $E(J) = B_{XY}J(J+1)$. The energy gap law suggests a correlation between the relaxation rates from the v_{XY} level and the v_π quantum number of the bending level with energy just below that of the v_{XH} state,

$$v_\pi = J_m = \mathrm{Int}(\sqrt{\Delta E \mathrm{vib}/B}) \approx \mathrm{Int}(\sqrt{\omega_{XH}/B}).$$

The rotational constants are of the order of 1.5–2 cm^{-1} for light diatomics XY and of 20 cm^{-1} for XH hydrides, so that J_m amounts to ∼30 for the former and to 14 for the latter. Because $B \sim 1/\mu$ and $\omega \sim \sqrt{1/\mu}$, $J_m \approx \omega/B \sim \sqrt{\mu}$, the J_m value increases on deuteration; for HCl and DCl, $J_m = 16$ and 19, respectively. The apparent breakdown of the $k = f(\Delta E)$ dependence is thus a direct consequence of the $k = f(\Delta v_{\mathrm{tot}})$ momentum gap relation.

The gap law suggests an exponential variation of the relaxation rates with J_m. Such a dependence,

$$\ln k_{\mathrm{vib}} \approx B - AJ_{\mathrm{m}}, \tag{4.16}$$

was established for hydrides and some other small molecules such as CH_3F [50] (Fig. 4.6). The theoretical treatment of the vibration–rotation coupling in slightly anisotropic potential is developed in [51-52].

4.3.3 The Vibrational Predissociation of Complexes of Polyatomic Molecules

The major part of rules deduced from the study of complexes of diatomics (the propensity to small Δv_{tot} and ΔE changes, a strong rotational excitation of dissociation fragments) are valid also for the polyatomics. So, the rotational energy ($E_{\mathrm{rot}} = 380$ cm^{-1}) of p-DFB fragments resulting from the $8^1 \to 0^1$ dissociation channel of the p-DFB·Ar complex is equal to 85% of the total energy [53].

Among the electronically excited complexes of aromatics, we choose the s-tetrazine-Ar (T.Ar) complex and aniline complexes, giving a good insight into the dissociation mechanisms.

The time-integrated emission spectra from different vibronic levels of the jet cooled ($T_{\mathrm{rot}} \approx 3$ K) T.Ar complex excited to the $^1B_{2u}$ state of tetrazine are composed of three groups of bands [54] (the symbols of levels and transition assigned to the complex are overlined). Upon the excitation of the $\overline{6a^1}$ level, the spectrum contains (Fig. 4.7)

- narrow bands emitted from the initial $\overline{6a}^1$ level with the ($\delta\omega \approx 2$ cm^{-1}) width limited by the spectral resolution,
- slightly broader ($\delta\omega \approx 3.7$ cm^{-1}) $\overline{16a_2^2}$ and $\overline{16a_1^1 16b_1^1}$ bands emitted from the $\overline{16a^2(\sigma^v)}$ and $\overline{16a^1 16b^1(\sigma^{v'})}$ levels (where σ^v and $\sigma^{v'}$ indicates excitation of v and v' quanta of an external mode,

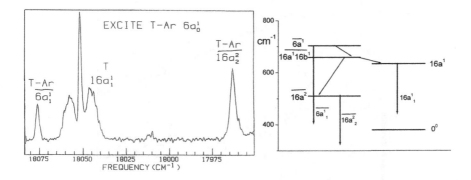

Fig. 4.7. (left) The fluorescence spectrum emitted upon the excitation of the tetrazine-argon complex to the $\overline{6a^1}$ level and (right) the scheme of relevant S_1 levels and transitions (from [54])

Table 4.6. Rate constants (in $10^8 s^{-1}$) of decay steps in T^*A dissociation

Level	k	Level	k	Level	k
$\overline{16a^2}$		$\overline{6a^1}$		$\overline{6b^2}$	
↓	4.6	↓	0.5	↓	16
$\overline{16a^1}$		$\overline{16a^1 16b^1}$		$\overline{16a^1 16b^1}$	
↓	9.5	↓	4.0	↓	?
0^0		$\overline{16a^2}$		$\overline{16b^2}$	
		↓	32	↓	22
		0^0		$\overline{16a^1}$	

- the $16a_1^1$ band from $16a^1$ level of the free molecule. Its contour shows an unresolved P and R subband structure corresponding to $J \leq 25$, indicating a large increase of the rotational temperature.

The time- and energy-resolved study of emission with a subnanosecond time resolution [55] shows that lower levels are populated by cascades from the higher ones. The analysis of the decay curves (Fig. 4.8) shows, for instance, that, upon the initial excitation of the $\overline{6b^2}$ level, the $\overline{16a^1 16b^1}$ state is directly populated by the $\overline{6b^2} \to \overline{16a^1 16b^1}$ transition but the decay of emission from the $\overline{16a^2}$ state cannot be fitted in this way. It corresponds to the stepwise $\overline{6b^2} \to \overline{16a^1 16b^1} \to \overline{16a^2}$ relaxation. Finally, the decay curve of the $16a^1$ emission of the free molecule indicates that this level is populated from the $\overline{16a^1}$ level of the complex. A few examples of such relaxation chains with the rates corresponding to each step are given in Table 4.6.

The relaxation proceeds by a step-by-step vibrational redistribution within the complex, with the energy transferred to external modes (E_σ) increasing in each step. The complex dissociates when E_σ exceeds the dissociation threshold of the complex of $D_0 \approx 330$ cm^{-1}, with the energy excess $E_\sigma - D_0$ transformed into the rotational

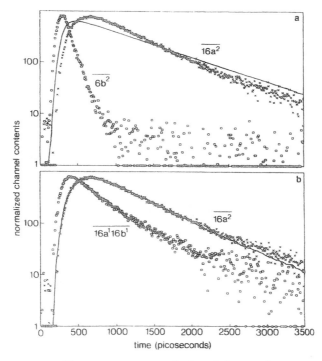

Fig. 4.8. Decay curves of fluorescence upon excitation of the tetrazine-argon complex to the $6b^2$ level (from [55])

energy of the polyatomic fragment. The energy remaining in the internal modes of $T^{\#}$ is a spectator of the dissociation process and is conserved in the free T fragment.

The rate-determining step in the relaxation of T.Ar complexes is the vibrational redistribution with the rate $k_{IVR} < k_{diss}$. This is not true for all systems: the relation between the k_{IVR} and k_{diss} rates depends on the nature of the ligand C in the $XY^{\#}$.C complex, as shown in the study of aniline complexes with Ar, N_2 and CH_4 (Table 4.6)) [56–58]. In the Ar complex, the lifetimes of initially excited vibronic levels are long and the energy excess is conserved in the free aniline molecule after dissociation, as in the case of tetrazine. In the CH_4 complex, the decay of the initial level is much more rapid and the free aniline molecule is formed uniquely in the vibrationless 0^0 state. The whole energy of aniline internal modes is transferred to external modes before the dissociation occurs so that dissociation and not IVR is, here, the rate-determining step. The kinetics of the N_2 complex is intermediary. The difference between Ar and CH_4 cases was related [58] to the difference in the density of vibrational levels of external modes—ρ_σ. The vibrational redistribution rate increases with ρ_σ, whereas the dissociation rate—treated in terms of RRKM theory—decreases. Because $\rho_\sigma(Ar) \ll \rho_\sigma(CH_4)$, $k_{IVR} < k_{diss}$ in the case of Ar and $k_{IVR} > k_{diss}$ in that of the methane complex.

Table 4.7. Decay rates (in 10^9 s^{-1}) and final levels F upon excitation of initial I levels of aniline*C complexes

C D_0 (cm^{-1})		Ar (450)		N_2 (500)		CH_4 (490)	
I	E_{vib}	$1/\tau$	F	$1/\tau$	F	$1/\tau$	F
$\overline{6a^1}$	(493)	1.4	0^0	22	$\overline{0^0}$	>100	$\overline{0^0, 0^0}$
$\overline{15^2}$	(718)	5.2	$0^0, 10b^1, 16a^1$	30	0^0	>100	0^0
$\overline{1^1}$	(803)	2.5	$0^0, 10b^1, 16a^1, I^1$	26	$0^0, 10b^1$	>100	0^0

The mechanism of the vibrational predissociation of complexes of large molecules is still not entirely elucidated. For instance, the energy distribution in the dissociation products of the benzene–rare-gas complexes was not completely explained [59]. The puzzling behavior of the p-difluorobenzene(DFB)–Ar complex [60–63] is briefly discussed in Sect. 4.2.2. The relation between the rates of IVR and dissociation steps may be more complex than assumed in the previous simple scheme.

Conclusion

The direct coupling of internal vibrational modes to the continua (dissociative continuum for XY.C complexes and quasi-continuum of phonons for XYC$_n$ active centers in solids) is so weak that, in the absence of other channels, the relaxation (predissociation) rates are negligibly small. A more efficient relaxation channel involves a sequential coupling of the internal mode to the external modes more strongly coupled to the continuum. The intracomplex (intracenter) vibrational redistribution seems to be the slow, rate-determining step at least in the case of rare-gas ligands or hosts.

4.3.4 The Vibrational Relaxation in Hydrogen-Bonded Systems

We will extend the discussion of relaxation/predissociation processes to a large class of polyatomic molecules containing the -X-H group involved in the hydrogen bond X-H..C. The major sources of information are infrared absorption spectra of liquids and fluid solutions recently extended to isolated gas-phase complexes or clusters. The potential energy of the X-H..C system may be approximated by a two-dimensional $V(r, R)$ surface where r_{XH} is the length of the X-H bond and R_{XC} indicates a set of coordinates of heavy X and C atoms. This surface is currently reduced to the one-dimensional potential curve assuming a complete separation of the rapid proton motion and of the slow displacements of heavy X and C atoms (atom groups) in a kind of Born–Oppenheimer approximation. A picture of a set of $U(R_{XH..C})$ potential curves corresponding to different v_{XH} vibrational states of the X-H oscillator (Fig. 4.9) will be applied for representation of the X-H..C vibrational spectra

Infrared Spectra

The specific feature of the infrared absorption spectra of hydrogen-bonded systems is a large frequency shift, $\Delta\omega$ −200 to −1,000 cm^{-1}, and width ($\delta\omega_{XH} \approx$ 100–300

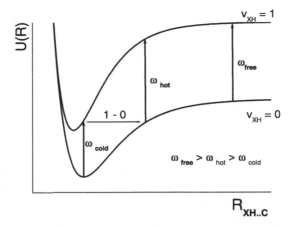

Fig. 4.9. A one-dimensional representations of the potential energy surface of a hydrogen-bonded system. The $v_{XH} = 0 \to 1$ radiative transitions and the $v_{XH} = 1 \to 0$ relaxation path are indicated

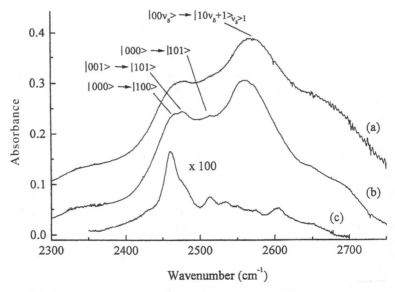

Fig. 4.10. Absorption spectra of the $(CH_3)_2O.HCl$ complex in the vapor, T = (a) 300 and (b) 230 K and (c) in a jet ($T_{vib} \approx 30$ K) (from [64])

cm^{-1}) of the band assigned to the $v_{XH} = 0 \to 1$ transition of X-H bound in the X-H..C hydrogen bond. These effects are larger by one order of magnitude or more than those observed for all other modes. The mystery of the exceptional broadening effects remained nonelucidated until recent studies of jet-cooled species and time-resolved experiments.

We will take as an example the collision-free $(CH_3)_2O$-HCl complex [64–65]. In the room-temperature gas, the narrow band of the free HCl molecule at 2,885 cm^{-1} is replaced by a complex broad-band structure with $\omega_{max} \approx 2{,}600$ cm^{-1} and the $\sim$2,700 and 2,470 cm^{-1} shoulders. This structure is strongly modified at lower temperatures: the spectrum of the jet-cooled complex ($T_{vib} \approx 30$ K) complex is composed of a relatively narrow origin band at 2,460 cm^{-1} followed by a series of partially resolved broader bands (Fig. 4.10). The contour of the origin band may be fitted by assuming $T_{rot} \approx 15$ K and the widths of individual rotational lines of 9.5 cm^{-1} for HCl and of 3 cm^{-1} for DCl.

The analysis of the temperature dependence of intensity distribution shows that the absorption spectrum is composed of overlapping bands corresponding to transitions between different $v_{XH} = 0$, $v_\sigma v_\pi$ and $v_{XH} = 1$, $v'_\sigma v''_\pi$ levels, where σ and π are X..C stretching and X-H..C bending modes. Because their frequencies ω_σ and ω_π are low (30–200 cm^{-1}), a large number of $v_\sigma \geq 1$ and $v_\pi \geq 1$ levels is populated at $T > 0$. Because the band structure cannot be completely resolved, the absorption spectra may be represented as transitions between different points of the $U(R_{XH..C})$ potential curve corresponding to different values of external mode energy $E_\sigma + E_\pi$.

The red shift of the $v_{XH} = 0 \to 1$ transition indicates that the bonding energy is larger and the equilibrium distance smaller in the $v_{XH} = 1$ than in the $v_{XH} = 0$ state (Fig. 4.9). The hot sequence bands from $v_\sigma \geq 1$ and $v_\pi \geq 1$ levels have thus higher frequencies than the cold $v_{XH} = 0 \to 1$ transition and tend with increasing v_s to the frequency of the free-molecule ω_{free}.

If the widths of individual rotational lines of the 0–1 transition are homogeneous, they correspond to the lifetime of the $v = 1$ level of 0.6 ps for HCl and of 1.85 ps for DCl. The agreement with the 0.9 ps relaxation time measured for this level of the complex in the CCl$_4$ solution [66] is excellent. The $\sim$10 cm^{-1} bandwidths seem to be characteristic for a wide class of 1:1 XH..C complexes with not very strong hydrogen bonds such as jet-cooled phenol complexes measured by the IR-UV double resonance technique [67–69]. These widths are consistent with decay times of the order of 0.5–1 ps measured in room-temperature liquids and solutions.

The exotic widths of X-H..C bands are thus due to a superposition of two effects:

- the large relaxation rates and homogeneous widths of the $v_{XH} = 1$ levels and
- a strong inhomogeneous broadening (sequence congestion) due to a dense set of levels involving internal and external low-frequency modes (ω_s, ω_b and so on) coupled to the X-H stretching.

Such a picture was already proposed in the early theoretical models of the X-H..C spectra [70, 71].

Relaxation and Predissociation

The conclusions from the absorption studies are confirmed by the real-time experiments, which show a significant increase of the relaxation rates of $v_{XH} = 1$ levels when the molecule is involved into a hydrogen-bonded complex. So the decay times

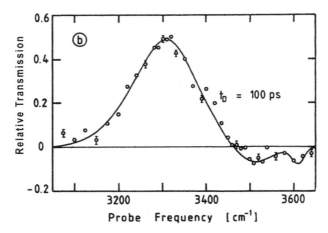

Fig. 4.11. The differential absorption spectrum of ethanol recorded with a 100-ps delay after the IR excitation at 3,300 cm^{-1} (from [76])

of free O-H groups of the methanol monomer ($\tau = 8$–10 ps) in the diluted CCl$_4$ solution [72] are reduced to the subpicosecond scale for methanol, forming a (CH$_3$OH)$_n$ cluster [73].

Such a high relaxation rate of $v_{XH} \geq 1$ levels of an X-H..C system is due to the strong coupling between the ($v_{XH} = 1$, $v_{XH..C} \approx 0$) and a set of ($v_{XH} = 0$, $v_{XH..C} \gg 1$) levels explained by an intersection between repulsive parts of $U_{v=1}(R_{XH..C})$ and $U_{v=0}(R_{XH..C})$ potential energy curves (Fig. 4.9) [74].

If the energy transferred to low-frequency external modes on relaxation of the $v_{XH} = 1$ level exceeds the X-H..C bonding energy, it will induce dissociation of the complex. Otherwise, this energy is conserved in the complex so that its vibrational temperature is increased. Such a temperature increase is detected for XH..C complex in solution upon a subpicosecond excitation by time-resolved absorption measurements. So the pulse excitation of the (CH$_3$)$_2$O-HCl complex in CCl$_4$ solution with $\omega_{exc} \approx 2{,}350$ cm^{-1}) produces a hole (reduced absorption) in the red part of its spectrum assigned to absorption of cold $v_{XH..C} \approx 0$ complexes. This hole is due to the population transfer from the $v_{XH} = 0$, $v_{XH..C} \approx 0$ to the $v_{XH} = 1$, $v_{XH..C} \approx 0$ level and decays with 0.9-ps decay time. Its decay is accompanied by the rise of the hot absorption at the blue edge of the band, which builds up with the same 0.9-ps rise-time and may be assigned to the $v_{XH} = 1$, $v_{XH..C} \approx 0 \rightarrow v_{XH} = 0$, $v_{XH..C} \gg 0$ nonradiative transition, creating an increased population of hot ground-state levels. The hot absorption decays with a 3.1-ps decay time [66].

The energy of 2,350 cm^{-1} transferred to external modes is not sufficient for dissociation of the strongly bound (CH$_3$)$_2$O-HCl complex. In contrast, the excitation of ethanol in an inert solvent with the energy of 3,300 cm^{-1} induces dissociation of (ROH)$_n$ oligomers. The hole formed in the 3,300 $\pm$ 100 cm^{-1} cold absorption region by a laser pulse decays with $\tau = 0.5$–1-ps decay time while the 3,500 and 3,600 cm^{-1} absorption bands build up with $t_{rise} \approx \tau$ and decay at a longer (2–10 ps) time scale

(Fig. 4.11). These bands correspond to the absorption of bound and free O-H groups in the $(ROH)_2$ linear dimer formed by dissociation of larger $(ROH)_n$ oligomers [73, 75–77].

4.4 The Competition Between Relaxation Channels

The vibrational relaxation of a polyatomic molecule excited to a high vibrational level, $|i\rangle$, is a stepwise $i \rightarrow j \rightarrow k$ descent on the ladder of lower vibrational levels. In most systems, the molecule has a choice between different $i \rightarrow j \rightarrow k$ and $i \rightarrow j' \rightarrow k$ paths. It would be important to learn whether this choice obeys some general rules or is determined in each case by intrinsic parameters of the molecule (such as the strength of the $i-j$ and $i-j'$ coupling) and of its environment. An abundant experimental material was discussed in review papers [9–10], but it is still difficult to conclude in spite (or because) of its abundance. We will not try to go to conclusions but only to show a little sample of data concerning a few selected systems.

4.4.1 Model Treatment

The recently developed theoretical techniques, such as the coupled-channel calculations (cf. Sect. 4.1.1), allow predicting the evolution of a given $XY^{\#} + C$ molecular system provided there is detailed knowledge of the potential and the initial state of the system. It is, however, difficult to generalize these results and establish the rules valid for a whole class of molecular systems. For this reason, simple schemes such as the Schwartz–Slavsky–Herzfeld (SSH) theory [78] "born in 1952 and still going strong", developed in the further works (SSH-T) [79–82] are widely used. The original SSH model describes a collision between two particles considered as isotropic (atom + diatom or diatom + diatom), and its extrapolation to polyatomic systems cannot be expected to give more than a purely qualitative description of relaxation processes.

In the SSH theory, the transitions take place in the two-dimensional X-Y..C system defined by the r_{X-Y} and $R_{XY..C}$ coordinates between the initial $|i\rangle = |v_i(r)\phi_i(R)\rangle$ and final $|f\rangle = |v_f(r)\phi_f(R)\rangle$ states, where $v(r)$ are vibrational functions of r and $\phi(R)$ are translational functions of the $XY + C$ pair. The rotational degrees of freedom are neglected.

The collisionally induced transition is expressed by the $V_{int}(r)V(R)$ perturbation, so that its probability of transition P_{if} is given by

$$P(i \rightarrow f) = \text{const}|\langle v_i|V_{int}(r)|v_f\rangle|^2 |\langle \phi_i(R)|V(R)|\phi_f(R)\rangle|^2, \qquad (4.17)$$

currently written in a simple form,

$$P(i \rightarrow f) = \text{const}|V_{if}|^2 I(\Delta E_{tr}), \qquad (4.18)$$

where V_{if} is the strength of the collision-induced coupling between vibrational levels, whereas the dependence of the transition probability on the amount of energy transferred to the continuum of translational energies is expressed by $I(\Delta E_{\mathrm{tr}})$. The V_{if} and $I(\Delta E_{\mathrm{tr}})$ factors cannot be easily calculated, but they reflect two aspects of the momentum-gap dependence of transition rates: a propensity to small changes of the vibrational quantum number, Δv, and of the linear momentum, Δk (i.e., of the translational energy $E_{\mathrm{tr}} \sim k^2$).

In the absence of detailed information, one can only use for the analysis of experimental data a drastically approximated form of Eq. 4.18 [22]. It is assumed that the V_{if} factor depends for all modes of each molecule on a single parameter: the overall difference between vibrational quantum numbers in the initial and final state, $\Delta v_{\mathrm{tot}} = \sum_k |v_k^i - v_k^f|$. The exponential dependence of V_{if} on Δv_{tot} is assumed to be

$$|V_{if}|^2 = A^{-\Delta v_{\mathrm{tot}}}, \tag{4.19}$$

where $A \approx 10$.

On the other hand, the $I(\Delta E)$ is approximated by

$$I(\Delta E) = \mathrm{e}^{-\alpha|\Delta \mathrm{E}|}, \tag{4.20}$$

where the value of $\alpha = 0.01$ was chosen for ΔE expressed in cm^{-1} units in order to fit experimental data for the whole family of methyl halides. A more realistic form of the $I(\Delta E)$ function is based on the Thompson theory of the $V - T$ relaxation in the Morse potential [83] and takes into account the reduced mass and relative velocity of the colliding pair. The $I(\Delta E)$ curves obtained by numerical integration show a maximum for finite ΔE value and tend to zero for $\Delta E \to 0$ and $\Delta E \to \infty$ [84]. For $\Delta E > 0$ (up-relaxation), $I(\Delta E)$ must be multiplied by the Boltzmann factor $\mathrm{e}^{-\Delta E/kT}$.

The formula expresses the propensity to small changes of vibrational quantum numbers and small energy transfers, unifying two aspects of the gap laws. The energy gap $\Delta E \approx \omega_i$ between v_i nd $v_i - 1$ levels of the same mode ($\Delta v_{\mathrm{tot}} = 1$) is usually larger than the $E_{v_i,v_j} - E_{v_i-1,v_j+1} \approx \omega_i - \omega_j$ gap so that the first factor in Eq. 4.21 favors the $i \to i - 1$ channel and the second one the $i, j \to i - 1, j + 1$ relaxation path.

$$P(i \to f) = \mathrm{const}|V_{if}|^2 \mathrm{e}^{-\alpha|\Delta E|} \tag{4.21}$$

4.4.2 Experiment

As indicated, we limit our presentation to a few results illustrating the competition between the $\Delta v_{\mathrm{tot}} \to 1$ and $\Delta E \to 0$ propensities for small molecules and a large diversity of behaviors not submitted to any clear rules for larger ones (benzene, p-DFB). We did not see any indication of a relation between the intramolecular coupling strength and the relaxation rate or of a propensity to the symmetry conservation.

Table 4.8. Energy gaps (cm^{-1}) and decay rates (s^{-1}) for HCN isotopomers in Xe matrices

Transition	H^{14}CN		H^{15}CN		DCN	
	(ΔE)	k	(ΔE)	k	(ΔE)	k
(001)→(110)	(477)	6.3×10^5	(508)	4.7×10^5	(208)	3.8×10^8
(110)→(040)	(−12.5)	eq	(−42)	eq	(240)	3.0×10^7
(040)→(030)	(691)	4×10^5	(688)	3.2×10^5	(540)	8.4×10^5
(100)→(030)	(−35)	eq	(−34)	eq	(204)	?
(030)→(020)	(701)	3.0×10^5	(700)	2.4×10^5	(575)	6.3×10^5
(020)→(010)	(706)	2.0×10^5	(706)	1.6×10^5	(563)	4.2×10^5
(010)→(000)	(717)	1.0×10^5	(716)	8.0×10^4	(573)	2.1×10^5

eq, pseudoequilibrium.

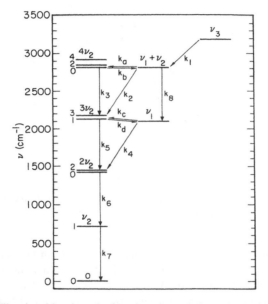

Fig. 4.12. Vibrational levels and relaxation channels for HCN in the Xe matrix

Small Molecules

Triatomics in Matrices

The relaxation of the $v_3 = 1$ (001) level ($\omega_3 = 3{,}311.47$ cm^{-1} in the gas) of HCN in a Xe matrix proceeds by a cascade involving the levels of $\omega_1 = 2{,}096.85$ cm^{-1} and $\omega_2 = 712.98$ cm^{-1} [85]. The scheme of transitions is given in Fig. 4.12 and their rates are listed in Table 4.8. The rates seem to be much more sensitive to energy gaps than to the Δv_{tot}: that of the (001) → (110) relaxation step with $\Delta v_{\text{tot}} = 3$ in HCN is of the same order of magnitude as those of the $\Delta v_2 = -1$ ($\Delta v_{\text{tot}} = 1$) steps with

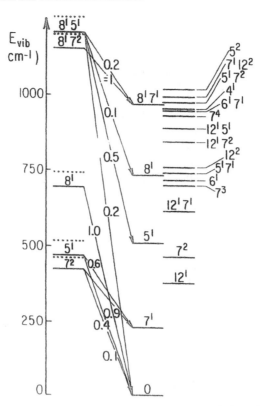

Fig. 4.13. The energy levels and transitions involved in the dissociation of excited S_1 levels of the jet-cooled glyoxal-H_2 complex (from [89])

larger energy gaps. The drastic enhancement of this rate in DCN and its reduction in $H^{15}CN$ is obviously related to differences in ΔE. Slightly increased rates of the $\Delta v_2 = -1$ steps in DCN may be explained in the same way. The propensity to small ΔE predominates over a tendency to minimize Δv_{tot} values. The same conclusions have been deduced from the study of CO_2 in rare-gas matrices [86].

Glyoxal

Electronically excited glyoxal ($C_2H_2O_2$) and glyoxal-d_2 have been extensively studied in gaseous mixtures, supersonic expansions and jet-cooled complexes. The number of efficient relaxation channels is limited and their choice seems to be governed by the propensity to small Δv_{tot} values. A high selectivity of relaxation paths is reported for low-energy glyoxal-helium collisions in a supersonic expansion. The 8^1 level decays only to the 7^1 and 0^0 states, whereas the $2^1 \rightarrow 0^0$ is the main decay channel from the high-energy ($E_{vib} = 1{,}391$ cm^{-1}) 2^1 state [87]. The same rules govern the vibrational predissociation of vibronic levels ($E_{vib} > 230$ cm^{-1}) of the weakly

Table 4.9. Branching ratios of transitions from the 6^1 level of benzene induced by collision with colliders C

C	0^0	$16^1/11^{1(a)}$	4^1	16^2	$6^1 16^1$
ΔE	-522	$-285/-7$	-157	-48	$+237$
		$T = 300$ K			
He	<0.025	0.17	[b]	<0.02	0.59
N_2	<0.05	0.22	[b]	<0.04	0.56
SF_6	0.115	0.26	[b]	0.145	0.26
i-C_5H_{12}	0.165	0.29	[b]	0.115	0.37
n-C_7H_{16}	0.15	0.20	[b]	0.13	0.35
		$T = 25$ K			
He	~0	0.62	~0	0.38	—
Ne	~0	0.80	~0	0.20	—
N_2	0.08	0.66	.	0.26	—
H_2	0.63	0.29	0.02	0.06	—
D_2	0.44	0.40	.	0.16	—
CH_4	0.36	0.53	.	0.11	—
c-C_3H_6	0.251	0.52	.	0.23	—
		He complex			
	0.09	0.20	.	0.71	—

[a] Unresolved emission from the 16^1 and 11^1 levels.
[b] The emission from 4^1 and 16^2 cannot be resolved at 300 K.

Table 4.10. Branching ratios in transitions from the 8^2 levels of p-DFB at 40 K for different collision partners C

Final level	0^0	30^1	8^1	30^2	$8^1 30^1$
ΔE (cm^{-1})	-360	-240	-186	-120	-65
Δ_{tot}	2	3	1	4	2
He	0.05	0.02	0.83	<0.01	0.10
Ar	0.07	0.11	0.48	0.05	0.29
Kr	0.18	0.12	0.36	0.05	0.28
H_2	0.13	0.01	0.79	0.01	0.06
N_2	0.23	0.07	0.46	0.02	0.16
CH_4	0.41	0.11	0.37	0.03	0.07
C_3H_8	0.61	0.07	0.17	0.02	0.05

bound ($D'_e \approx 40$ cm^{-1}) glyoxal-H_2 complex (Fig. 4.13) [88–90]. All efficient relaxation channels correspond to $\Delta v_{tot} = 1$ or 2. The dissociation patterns depend on the structure of the complex. Two isomeric forms, C and C$'$, of the glyoxal Ar and Kr complexes [89] have different dissociation channels [90]. In the room-temperature vapors, the collisions with the ground-state glyoxal molecules induce transitions from the 8^1 level with ($E_{vib} = 735$ cm^{-1}) to 4 among 15 energetically accessible levels.

The energy gaps are $+233, -17, -502$ and -735 cm^{-1}, respectively, for transitions to $8^1 7^1, 6^1, 7^1$ and 0^0 levels. The closely spaced 5^1 and 7^2 levels with $\Delta E = -226$ and -269 cm^{-1} are not populated. Note that ω_7 is the lowest frequency mode and that 6, 7 and 8 modes correspond with out-of-plane vibrations [14].

Benzene and Its Derivatives

We have at our disposal an enormous amount of data concerning the collisional relaxation in room-temperature gases and supersonic expansions and the vibrational predissociation of complexes of small aromatic molecules in the S_1 state: benzene [22, 91, 92], aniline [23, 58], p-difluorobenzene (p-DFB) [24, 60–63, 93, 94], etc. The selectivity and propensity to small Δv_{tot} and small ΔE_{vib} factors is present. It seems, however, difficult to establish the rules valid for a variety of molecular systems because the choice of relaxation channels varies with the collision partner and temperature (average energy of collisions).

In benzene, the branching ratios for transitions from the 6^1 level of the S_1 state (Table 4.9) show a pronounced dependence on the collision partner. The efficient up-relaxation observed in room-temperature gas is, of course, suppressed when the temperature is lowered. For rare-gas atoms and N_2 quasi-atom, the transitions to the close-lying 16^2 state are favored in spite of the high value of $\Delta v_{tot} = 3$. The high probability of $6^1 \rightarrow 0^0$ transition in collisions with H_2, D_2 and CH_4, in which the $V \rightarrow V$ energy transfer is impossible, may be tentatively explained by the vibration-to-rotation energy transfer. The large values of the rotational constants of energy acceptors allow a transfer of large quantities of energy ΔE values for small ΔJ. On the other hand, the transfer of larger amounts of energy in collisions with heavier partners may be explained either by the $V \rightarrow V$ transfer or by statistical energy redistribution in long-lived collisional complexes.

In p-DFB (Table 4.10), the propensity to $\Delta v_{tot} = 1$ is pronounced for light collision partners (He, H_2), whereas the effects of heavier ones are much less selective. The molecules induce more efficiently the $8^2 \rightarrow 0^0$ transition with $\Delta E = 360$ cm^{-1} than rare gases, although the $V \rightarrow V$ transfer is impossible in the case of N_2 and methane. On the other hand, it seems difficult to explain the high efficiency of CH_4 collisions for the $8^2 \rightarrow 0^0$ relaxation by the $V \rightarrow R$ transfer, as in the case of benzene, in view of the low efficiency of H_2-DFB collisions.

Puzzling effects were observed in predissociation of p-DFB complexes [60]. Upon a selective excitation of two closely spaced $\overline{5^1}$ and $\overline{6^2}$ levels with energies $E_{vib} = 817$ and 820 cm^{-1} exceeding by ~ 400 cm^{-1} the dissociation threshold, the free DFB is formed in the 0^0 and in nearly resonant 6^1 state but the 0^0 to 6^1 branching ratios are of ~ 1 in the first and of ~ 5 in the latter case and the decay rates of levels are different by a factor of 5.

Conclusions

The question about the factors determining relative efficiencies of different relaxation channels is still open.

There is no obvious relation between the intrinsic coupling strength and the probability of the collision-induced transition for different pairs of vibrational levels.

The propensity to small $\Delta E(\Delta k)$ and Δv_{tot} values plays certainly an important role in the choice of relaxation pathways so that the

$$k_{rel} \sim e^{-A\Delta v_{tot}} e^{-\alpha |\Delta E|}$$

relation is useful as the zero-order approximation. The relative importance of the Δv_{tot} and ΔE factors seems to change from case to case, and there is no reason to believe that the same values of A and α parameters may be applied to different molecular systems. The problem is complicated by the pronounced dependence of branching ratios on the collision partner and the collision energy. The SSH theory, in spite of its local successes, is not able to give a quantitative description of relaxation mechanisms.

It seems to us that a complete theoretical treatment of a well-chosen $XY^{\#} + C$ collision and of an $XY^{\#}C$ complex is necessary. It must be associated to the experimental study of the $XY^{\#} + C$ collisions and $XY^{\#}C$ complexes with a wide distribution of collision partners, collision energies and predissociation conditions. A good choice of molecular systems with well-known intramolecular parameters is fundamental. Small polyatomic molecules, such as HCN, CH_2O and C_2H_2, seem to be good candidates.

5

Electronic Relaxation

5.1 Potential-Energy Surfaces

The electronic relaxation is a spontaneous or medium-induced nonradiative transition between vibronic (ro-vibronic) levels of two different electronic states, i.e., between two adiabatic potential-energy surfaces in contrast with previously discussed phenomena (vibrational redistribution and relaxation) that take place on a single potential-energy surface.

It is natural to consider at the outset the energies and wavefunctions of the electrons for fixed nuclei, because due to the vastly different masses of nuclei and electrons the nuclear kinetic energy term of the Hamiltonian, $-\sum_k (\hbar^2/2M_k)\nabla_k^2$ (where k numbers the nuclei and M_k are the masses of the nuclei), is much smaller than the electronic energy term $-\sum_i (\hbar^2/2m)\nabla_i^2$ (where i numbers the electrons and m is the electron mass). The *adiabatic* potential-energy surface of the electronic state A is the ensemble of clamped-nuclei energies, $U_A(Q)$, for all nuclear geometries $\{Q\}$. These are the eigenvalues of the electronic Hamiltonian, H_e, which is the total Hamiltonian minus the nuclear kinetic-energy part,

$$H_e \psi_A(q, Q) = U_A(Q)\psi_A(q, Q). \tag{5.1}$$

Here, the indices $A, B, \ldots$ label the electronic states of the molecule. The electronic wave functions, $\psi(Q, q)$, depend on the electronic coordinates q, but they also vary parametrically with the geometry Q.

The exact ro-vibronic wave function $\Psi(q, Q)$ of the system is the solution of the full molecular Hamiltonian,

$$(H_e + H_{vr})\Psi = E\Psi. \tag{5.2}$$

Because $H_{vr}(Q)$ is small, it makes sense to expand the full wave function in terms of the electronic eigenfunctions,

$$\Psi(q, Q) = \sum_B \psi_B(q, Q)\phi^{(B)}(Q), \tag{5.3}$$

for each nuclear configuration Q. We will show that the geometry-dependent co-efficients $\phi^{(B)}(Q)$ turn out to be the rotational-vibrational wave functions. For the sake of simplicity, in what follows, we will consider only a single nuclear coordinate (massweighted normal mode) Q. We insert the (formally) exact expansion (5.3) into the full Schrödinger equation (5.2). Using the fact that, according to (5.1), ψ_B are eigenfunctions of H_e, and taking into account the fact that the ro-vibrational (nuclear kinetic energy) part of the Hamiltonian $H_{vr}(Q)$ acts on the electronic wave functions $\psi(q, Q)$ because of its dependence on Q, we obtain a system of coupled differential equations describing the motion of the nuclei in an electronic state A, coupled with the nuclear motion in all other electronic states B,

$$\left\{ H_{vr} + U_A(Q) - E \right\} \phi^{(A)}(Q) = - \sum_B \langle \psi_A | H_{vr} | \psi_B \rangle \phi^{(B)}(Q). \qquad (5.4)$$

The integration in the brackets $\langle \cdots \rangle$ in (5.4) is taken over all electronic coordinates. Equation (5.4) is formally exact.

The right-hand side of Eq. (5.4) may be neglected when the electronic wave functions depend only weakly on Q and hence $H_{vr}(Q)\psi_A(q, Q)$ is small. In that event, we obtain the Born-Oppenheimer (BO) approximation. The sums in Eqs. (5.3) and (5.4) now reduce to a single term:

$$\Psi \approx \psi_A(q, Q)\phi^{(A)}(Q), \qquad (5.5)$$

and the left-hand side of Eq. (5.4) is the rotational-vibrational Schrödinger equation for each electronic state A. $\phi^{(A)}(Q)$ is the rotational-vibrational factor of the total wave function in any given level associated with the electronic state A, and is itself the product of the rotational wave function and the vibrational factor $\chi_{v_A}(Q)$. The factorization of the total wave function into an electronic and a rotational-vibrational part is an essential aspect of the BO approximation. These states will be the zero-order states in the remainder of this chapter. The BO approximation is good for rovibronic levels that are energetically well separated from the levels of other electronic states, which is the case basically with electronically non-degenerate ground states. Within the BO approximation the nuclear motion in the electronic state A is determined uniquely by the electronic potential $U_A(Q)$, which is strictly defined and can, at least in principle, be calculated with the analytical and numerical tools of *quantum chemistry*. No detailed knowledge (beyond symmetry considerations) of the electronic wave function ψ_A is required: this is probably the most powerful aspect of the BO approximation, which made possible the spectacular progress of the knowledge of molecules from the 1950's to the 1980's, before the advent of powerful computers and quantum-chemical codes.

The coupling terms on the right-hand side of Eq. (5.4) formally contain all of the dynamic coupling of electronic states, and in the quantum mechanical description they are thus the source of electronic relaxation phenomena which are the topic of this chapter. The coupling terms $\langle \psi_B | H_{vr} | \psi_A \rangle$ are electronic quantities, which are geometry-dependent (sometimes very strongly as we shall see below). They contain contributions due to both the vibrational and the rotational part of the nuclear kinetic energy operator. A detailed derivation of the molecular Hamiltonian shows that they also contain small contributions due to the fact that the translational motion of the molecule is separated. Because H_{vr} is a second-order differential operator, the noncommutation of $H_e(q, Q)$ and $H_{vr}(Q)$ will give cross-terms of the form

$$-\hbar^2 \langle \psi_B | \nabla_Q | \psi_A \rangle \nabla_Q \phi^{(A)}(Q), \tag{5.6}$$

as well as terms

$$-\frac{\hbar^2}{2} \langle \psi_B | \nabla_Q^2 | \psi_A \rangle \phi^{(A)}(Q). \tag{5.7}$$

The form of Eqs. (5.6) and (5.7) confirms that the BO approximation is recovered when the variation of the electronic wave functions with Q is so small that $\nabla_Q \psi$ can be neglected. The terms (5.6) are often thought to be more important than the terms in Eq. (5.7) and they are primarily considered when electronic relaxation is discussed. This coupling mechanism will be called *non-Born–Oppenheimer* or *vibronic coupling*. In fact these terms account also for the electronic-rotational Coriolis coupling. This mechanism does not couple the states belonging to different spin manifolds (as e.g., singlet and triplet), which are coupled by the spin–orbit interaction, i.e., the relativistic part of the electronic Hamiltonian.

The electronic factors in (5.6) and (5.7) become large when the electronic wave function changes its character rapidly as the nuclear geometry is varied adiabatically. This occurs, in particular, near avoided crossings of two surfaces $U_A(Q)$ and $U_B(Q)$. In diatomic molecules the non-crossing rule is strict and forbids the crossing of potential curves of the same symmetry. Therefore, avoided crossings such as the ones represented in Fig. 5.1. occur frequently. In polyatomic molecules the non-crossing rule is relaxed, but still plays a role, e.g., near conical intersections.

Probably the best-known examples of avoided crossings in diatomic molecules are those of the covalent and ionic potential-energy curves occurring in alkali-halides. In some of these compounds the avoided crossings are quite weak, which amounts to a quite abrupt change of the electronic wave function, as one follows the upper or lower adiabatic branch, and the electronic coupling term $\langle \psi_B | \nabla_Q | \psi_A \rangle$ approaches the form of a δ function. The coupled equations (5.4) remain correct of course in this situation, but they offer little physical insight and are numerically and analytically difficult to handle. It makes more physical sense to connect the curves *diabatically* across the anticrossing, because then, e.g., the *upper* covalent branch of the pair of states to the left of the avoided crossing is physically meaningfully connected to the *lower* branch to the right of the crossing which is also covalent. One is thus led to consider *diabatic* potential curves or surfaces, as a physically more satisfying alternative to the adiabatic surfaces defined by (5.1). We number them $a, b, \dots$ in order to distinguish them from the adiabatic states and surfaces.

The question that arises now, is how can one define the diabatic curves (the dotted lines in Fig. 5.1). Clearly, unlike the adiabatic curves, they do not correspond to the eigenvalues of the complete electronic Hamiltonian, H_e (cf. Eq. (5.1)). Instead H_e is partitioned according to $H_e = H_e^{(0)} + H_e^{(1)}$. This partitioning is a priori arbitrary, but we postulate that $H_e^{(1)}$ must be appropriately defined so that the (crossing) diabatic states $\psi_a^{(diab)}(q, Q)$ are solutions of the clamped-nuclei electronic Schrödinger

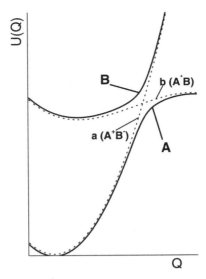

Fig. 5.1. A schematic representation of the crossing of ionic and covalent potential curves of alkali halides

equation with $H_e^{(0)}$ instead of H_e, i.e., $H_e^{(0)}\psi_a^{(diab)}(q, Q) = U_a^{(diab)}(Q)\psi_a^{(diab)}(q, Q)$, which replaces Eq. (5.1). We shall discuss later how this can be achieved in practice. The coupled equations (5.4) are obtained again in exactly the same form, but with H_{vr} replaced by $H_e^{(1)} + H_{vr}$ (and with the indices A and B appropriately replaced by a and b and also replacing $\psi_A(q, Q)$ by $\psi_a^{(diab)}(q, Q)$). Ro-vibronic wave functions consisting of a single product analogous to Eq. (5.5) constitute an alternative set of diabatic zero-order states. In addition to the dynamical coupling terms analogous to Eqs. (5.6) and (5.7), there will now be electronic contributions to the coupling of the form

$$V_{ab}(Q)\phi^{(a)}(Q) = \langle \psi_b^{(diab)}(q, Q)|H_e^{(1)}(q, Q)|\psi_a^{(diab)}(q, Q)\rangle\phi^{(a)}(Q), \qquad (5.8)$$

arising from the fact that part of the electronic Hamiltonian has been left out in the definition of the diabatic states and surfaces.

Returning to the example of covalent-ionic interactions in alkali halides, we may intuitively apprehend the usefulness of the diabatic states by considering Fig. 5.1. If a diabatic representation is chosen, the covalent (or ionic) character of an electronic state is preserved across the crossing zone and the abrupt change of the electronic wave function near Q_c is suppressed. Therefore, we may expect that the dynamical terms Eqs. (5.6) and (5.7) (which tend to become huge near Q_c in the adiabatic picture) are small or even zero. Instead, the electronic perturbation $V_{ab}(Q) = \langle \psi_b^{(diab)}|H_e^{(1)}|\psi_a^{(diab)}\rangle(Q)$ must be taken into account. Because $H_e^{(0)} + H_e^{(1)}$ produce the avoided crossing of U_A and U_B near $Q \approx Q_c$, while $H_e^{(0)}$ is defined to produce the crossing curves U_a and U_b, it follows that at the crossing

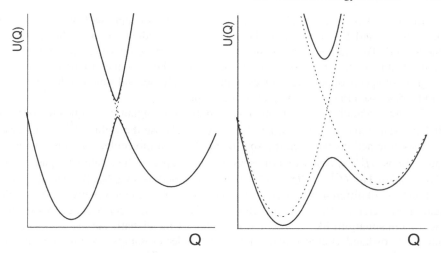

Fig. 5.2. Potential-energy curves for small (left) and large (right) values of the coupling constant V_{ab}

$|U_A(Q_c) - U_B(Q_c| \approx 2V_{ab}(Q_c)$. Thus, we see that the diabatic picture is preferable when the electronic coupling $V_{ab}(Q_c)$ is smaller than the dynamical coupling terms (5.6) and (5.7) would have been in the adiabatic picture.

In general, the dynamical properties of a molecule depend on the effective strength of the interaction between electronic states. In view of the preceding discussion, we may distinguish two limiting cases of the surface crossing as well as an intermediate case.(Fig. 5.2):

1) The electronic interaction V_{ab} is so large in an extended range of Q-values around the intersection point that the adiabatic surfaces $U_A(Q)$ and $U_B(Q)$ are well separated. These strongly avoided crossings may cause peculiar shapes of potential surfaces, such as energy barriers separating energy minima, or "shelf states." In this case, the adiabatic states ψ_A and ψ_B are slowly varying function of Q, so that the dynamical coupling is small and the BO approximation remains valid. The residual dynamical coupling A–B may be considered as a weak perturbation.

2) When $V_{ab} \to 0$, the variation of ψ_A and ψ_B with Q is so rapid in the vicinity of the crossing point that the coupling $\nabla_Q \psi_A$ and $\nabla_Q^2 \psi_A$ becomes large. This corresponds to the breakdown of the BO approximation, in the strict sense, implying a high probability of transitions $A \leftrightarrow B$. In a diabatic picture, however, based on crossing surfaces, the effective coupling a–b is electronic rather than dynamical, and will be much smaller than the coupling would be if the formal BO procedure was followed strictly.

3) An intermediate situation occurs when neither the adiabatic nor the diabatic approach is appropriate. The energy exchange between the nuclear and electronic degrees of freedom is the largest in this intermediate case.

In the first part of this chapter, we will treat the systems with well-separated surfaces U_A and U_B, which have been extensively studied and are relatively well understood. The studies of the systems involving surface crossings and giving rise to ultrafast relaxation processes have been developed during the last 10 years due to progress of spectroscopic techniques with the subpicosecond time resolution. They will be discussed in the second part of the chapter.

Before concluding this introduction we return once again to the question of how diabatic states may be defined, i.e., how can we choose the right $H_e^{(1)}$. In the case of weak interactions between the spin manifolds of a molecule, e.g., singlet-triplet interactions, $H_e^{(1)}$ is easily identified as corresponding to the relativistic part of the Hamiltonian, i.e., $H_e^{(1)} = H_e^{(rel)}$. $H_e^{(0)}$ then corresponds to a well-defined total spin S, and surfaces of different S may cross. In other cases of interactions between electronic states however, it is often more difficult to precisely define diabatic states, despite the fact that in situations like that depicted by Fig. 5.1 the diabatic curves can be drawn intuitively by hand. A clue as to the nature of avoided crossings is afforded by the example of ionic-covalent interactions mentioned above. This interaction couples states of clearly identifiable electronic configurations (ionic and covalent, respectively) so that we may conjecture that the interaction V_{ab} between the diabatic configuration states is due to the configuration interaction. In *ab initio* calculations the electronic wave functions $\psi_A(Q)$ are typically obtained in a configuration interaction (CI) treatment. They are represented as superpositions of electronic configurations of the form

$$\psi_A(Q, q) = \sum_k a_{Ak}(Q) f_k(q, Q), \tag{5.9}$$

where the a_{Ak} are Q-dependent configuration mixing coefficients. In well-defined situations as the ionic-covalent interaction example cited here, we may assume that each configuration state $f_k(q, Q)$ evolves slowly with nuclear geometry Q so that

$$\psi_A(Q, q) \approx \sum_k a_{Ak}(Q) f_k(q). \tag{5.10}$$

In this event the dynamical coupling terms (5.6) and (5.7) are due uniquely to the strongly Q-dependent configuration mixing coefficients. Conversely, the diabatic states correspond to configuration states with constant coefficients a_{ak}. Thus, we see that, in this situation, the electronic Hamiltonian term $H_e^{(1)}$ is related to the interelectronic repulsion $1/r_{ij}$ that causes the configuration interaction, and we can say that the avoided crossing is due to the covalent-ionic configuration interaction. It should be stressed, however, that this concept, while physically quite appealing and quite useful in qualitative discussions, becomes often blurred as soon as actual quantitative calculations of molecular dynamics are attempted.

5.2 Nonintersecting Potential-Energy Surfaces

In a large number of molecular systems, the potential-energy surface is not significantly modified by the electronic excitation. For instance, the equilibrium configuration of a large aromatic hydrocarbon molecule is only slightly changed when a single π electron is transferred from the bonding to the antibonding orbital. The surfaces do not intersect in the relevant part of the configuration space. The vibrationless level $v_B = 0$ of the upper electronic state, B, is quasi-isoenergetic with high vibrational levels, $v_A \gg 0$, of the lower electronic state, A.

The electronic relaxation processes in an isolated molecule are determined by the coupling strength $V_{v_A v_B}$ between these levels and the spacing $\Delta E_{AA'}$ (expressed often in terms of the density ρ_A of levels v_A situated near $v_B = 0$). Generalizing Eq. (5.6) to a multimode system we write the vibronic coupling between individual vibrational levels as

$$V_{v_A v_B} = -\hbar^2 \sum_k \left[\langle \psi_A | \partial/\partial Q_k | \psi_B \rangle \, \langle \chi_{v_k}^{(A)}(Q_k) | \partial/\partial Q_k | \chi_{v_k'}^{(B)}(Q_k) \rangle \right. $$
$$\left. \times \prod_{i \neq k} \langle \chi_{v_i}^{(A)}(Q_i) | \chi_{v_i'}^{(B)}(Q_i) \rangle \right], \qquad (5.11)$$

where $\chi_{v_i}^{(A/B)}(Q_i)$ are the vibrational wave functions in the electronic state A or B associated with the vibrational mode i, and the first integration $\langle \ldots \rangle$ is over the electronic coordinates while the second and third integrations are over ro-vibrational coordinates. In real situations one term in the $\sum_k$ often dominates, i.e., there is a single "promoting mode," so that the sum in Eq. (5.11) can be restricted to that term. If one considers spin-orbit interaction between states of different S (intersystem crossing, or, more generally, if a diabatic approach is used), the coupling takes the form

$$V_{v_a v_b} = \langle \psi_a | H_e^{(\text{rel})} | \psi_b \rangle \prod_i \langle \chi_{v_i}^{(a)}(Q_i) | \chi_{v_i}^{(b)}(Q_i) \rangle \qquad (5.12)$$

assuming that the spin-orbit coupling ($H_e^{(\text{rel})}$) or the relevant configuration interaction term does not vary significantly with Q.

The energy gap between the minima of the U_A and U_B surfaces is large as compared with the average frequency of the ground-state vibrational modes,

$$\Delta U_{AB} = U_B(Q_{B\,\min}) - U_A(Q_{A\,\min}) \gg \langle \omega_A \rangle.$$

Typical aromatic molecules have $\Delta U_{AB} \approx 20,000$ to $30,000 \, \text{cm}^{-1}$ and $\langle \omega_A \rangle \approx 1,000$ cm^{-1}. The v_A levels isoenergetic with $v_B = 0$ are high overtones and combinations of ω_A^i modes. The overall difference between the quantum numbers of both states, $\Delta v_{\text{tot}} = \sum_i v_A^i$, is large, of the order of $\Delta U_{AB}/\langle \omega_A \rangle$, and increases with increasing energy gap ΔU_{AB}.

The decay paths of the v_B level of an isolated molecule depend on the nature of the manifold of v_A. In a small molecule, v_B is the bright $|s\rangle$ state coupled to one or a few discrete levels, v_A ($|\ell\rangle$), giving rise to a set of φ states sharing the radiative properties of the bright state. On the other hand, the dense set of the v_A levels of a large molecule may be approximated by a dissipative continuum $\{m\}$, and the decay of the level $v_B = 0$ may be treated in terms of the large-molecule limit.

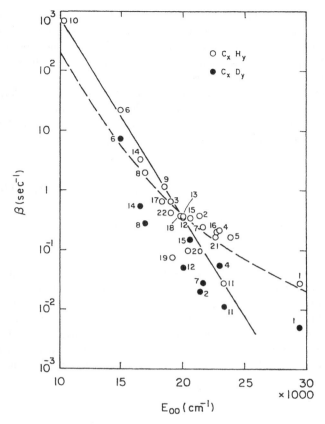

Fig. 5.3. Nonradiative decay rates $\beta \equiv k_{ISC}$ plotted vs. $T_1 - S_0$ energy gap $E_{00} \equiv U_{AB}$ for aromatic hydrogenated and deuterated hydrocarbons (for numbers, cf. [2])

5.2.1 Large Molecules

We choose the polynuclear aromatic hydrocarbons as model systems. The decays of their excited singlet S_1 and triplet T_1 states are practically insensitive to environment effects so that the results of early phosphorescence and fluorescence studies in rigid solutions allowed formulating simple rules concerning the rates of the $T_1 \rightarrow S_0$ intersystem crossing (ISC), $S_1 \rightarrow S_0$ and $S_{i>1} \rightarrow S_1$ internal conversion (IC) processes which are valid also for free molecules [1].

Relaxation of T_1 States

The rate constants of the nonradiative decay $T_1 \rightarrow S_0$ decay $k_{ISC} = (1 - Q_{ph})/\tau_{ph}$, deduced from the phosphorescence lifetimes (τ_{ph}), yield (Q_{ph}) in rigid organic glasses (for review, cf. [3]), show a pronounced dependence of k_{ISC} on the energy gap between T_1 and S_0 states ΔU_{ab} (Fig. 5.3), which may be approximated by

Table 5.1. Deuterium effect on the decay rate of T_1 states of aromatic hydrocarbons (k_{rel}) (from [2])

Molecule	Formula	ΔU_{AB}	k_{rel} s^{-1}
Benzene	C_6H_6	29,500	0.029
	C_6D_6		0.005
Biphenyl	$C_{12}H_{10}$	22,900	0.22
	$C_{12}D_{10}$		0.057
Naphthalene	$C_{10}H_8$	21,300	0.39
	$C_{10}D_8$		0.02
Pyrene	$C_{16}H_{10}$	16,900	2.0
	$C_{16}D_{10}$		0.29
Anthracene	$C_{14}H_{10}$	14,900	22
	$C_{14}D_{10}$		7.1

$$\ln k_{ISC} \approx k_0 - a\Delta U_{ab}. \tag{5.13}$$

This relation is valid also for the deuterated species but with much smaller k_0 value [1, 2] (Table 5.1).

These effects are not due to a different strength of the spin-orbit coupling as evidenced by the radiative decay rates roughly the same for all compounds but must be explained by the *energy gap* dependence of the vibrational factor $\langle \chi_{v_a} | \chi_{v_b} \rangle$ on ΔU_{ab}.

In the first approximation, this dependence may be described in terms of the large-molecule-limit scheme. It is assumed that the v_a levels with $E_{vib} \approx \Delta U_{ab}$ are so strongly mixed that they give rise to a set of quasi-identical levels with the $V_{v_a v_b}$ coupling constant, which depends only on their energy, i.e., on the average vibrational quantum number $\langle v \rangle = \Delta U_{ab} / \langle \omega_a \rangle$,

$$V_{v_a v_b} \sim e^{-a'\langle v \rangle}. \tag{5.14}$$

This value introduced into the Fermi Golden Rule gives

$$k_{ISC} = |V_{v_a v_b}|^2 \rho_a = k_0 e^{-a\langle v \rangle} \rho_a = k_0 e^{-\alpha \Delta U_{ab}/\langle \omega_a \rangle} \rho_a \tag{5.15}$$

equivalent to Eq. 5.13.

This model describes, in a correct way, the energy gap dependence of the relaxation rates but leaves unexplained two important points,

- The observed decay rates do not depend on the a-level density ρ_a, which increases rapidly with ΔU_{ab} and is modified on the alkyl substitution of the hydrocarbon molecule.
- The deuterium effect is too large to be explained by a change of the factor $\langle \omega_a \rangle$ in (5.15) resulting from the reduced frequency of the CH/CD vibrations.

These features may be explained in the framework of the *sequential-coupling* scheme assuming that the $|s\rangle = |v_b = 0\rangle$ state is efficiently coupled to a limited set of ground-state levels $|\ell\rangle$, which are, in turn, strongly coupled to a dense $\{m\}$ quasi-continuum of all other v_a levels (cf. Sect. 1.3.3).

We identify the ℓ-levels as overtones of the $\omega_{\text{C-H}}$ stretching modes v_{CH} and the m-levels as combinations of ring modes ω_{ring}. The frequencies of C-H modes are of the order of 3,000 cm^{-1}, whereas those of ring modes do not exceed 1,600 cm^{-1} so that, in a molecule with the energy gap $\Delta U_{ab} \approx 25,000$ cm^{-1}, the ground-state levels quasi-isoenergetic with $v_b = 0$ are $v_{\text{CH}} = 8$ and $v_{\text{ring}} \geq 15$. In view of the Δv dependence of the coupling strength, the values of $V_{s\ell}$ are significant only for the levels involving high-frequency vibrations. This coupling is strongly reduced by the deuterium substitution because $\omega_{CD} \approx 2,150$ cm^{-1} implies $v_{CD} \approx 11$.

The electronic relaxation of the T_1 states is thus a sequential process with the electronic energy transfer to the active, energy accepting ground-state modes as the first rate-determining step. This energy is then redistributed among a large number of modes with the $\sim 10^{12}$ s^{-1} IVR rate. The overall decay rate, determined by that of the slow $s \to \ell$ step, is independent of the $V_{\ell m}$ coupling strength and of the overall level density ρ_m. This rate is not significantly modified either by a rapid vibrational relaxation of the high vibrational ground-state m-levels $S_0^{\#}$ in condensed phases. As discussed in Sect. 1.4.3, the $k_{s\ell} = V_{s\ell}^2 \rho_\ell$ rates remain unchanged when the ℓ-states are diluted in the $\{m\}$ continuum.

The Energy-Gap Law and Kasha Rule

The energy-gap dependence of the internal conversion (IC) from the $S_1(v = 0)$ singlet level of aromatic hydrocarbons is not as well documented as the T_1 state decay. k_{IC} cannot be directly measured but is only estimated from the difference between more directly measured rate constants: $k_{\text{IC}} = k_s - (k_{\text{rad}} + k_{\text{ISC}})$ and represents usually a minor decay channel. The general tendency is nevertheless evident, $k_{\text{IC}} \approx 0$ for benzene and its derivatives with $E_{S_1-S_0} \approx 40,000$ cm^{-1}, whereas the IC rate increases at each step in the series naphthalene-anthracene-tetracene–pentacene and attains 10^8 s^{-1} for pentacene ($E_{S_1-S_0} \approx 17,300$ cm^{-1}).

The energy gap dependence of k_{IC} is the reason of a striking difference between the slow relaxation of S_1 and of the ultrafast decay of higher $S_{i>1}$ singlet levels. The empirical Kasha rule [4] deduced from early studies of luminescence in condensed phases stated that, upon excitation to singlet (S_i) states, only the emission from the lowest (S_1 and/or T_1) levels occurs, the $S_{i>1}$ state population being so rapidly relaxed that the $S_i \to S_0$ fluorescence is entirely quenched.

The Kasha rule is valid for an overwhelming majority of aromatic and aza-aromatic molecules not only in solutions but also in collision-free conditions. The first observed exception from the Kasha rule [5]: the $S_2 \to S_0$ emission of azulene at 28,200 cm^{-1} with a nonnegligible quantum yield ($Q_f \approx 0.05$ in solution [6]) and decay time $\tau = 3.2$ ns [7] in a supersonic jet) was considered as an isolated exotic case.

A few other examples of emission from higher singlet states were reported later:

- The $S_2\,^1A_1 \rightarrow S_0\,^1A_1$ fluorescence of thiophosgene Cl_2CS ($\omega_0 = 34,278\,cm^{-1}$) with a $\sim$30 ns lifetime and $Q_f \approx 1$ quantum yield, the internal conversion from the S_2 to S_1 state at $18,716\,cm^{-1}$ being practically absent [8–10].
- The weak and short-lived emission from the S_2 state at $23\,500\,cm^{-1}$ state of xanthione competing with an efficient relaxation to the S_1 ($E = 15,100\,cm^{-1}$) state [11–13].

The common feature of these systems is an unusually large energy gap between the S_2 and S_1 states. The Kasha rule breaks down in the case of azulene and thiophosgene, with nearly identical $S_2 - S_1$ and $S_1 - S_0$ gaps, whereas the weak S_2 emission of xanthione corresponds to the $S_2 - S_1$ gap, which is large but smaller than the $S_1 - S_0$ one. The Kasha rule is thus a direct consequence of the energy gap law applied to molecules with an even number of electrons; in this system, the $S_i - S_{i-1} (i \geq 2)$ energy gaps between the higher excited states are small as compared with the $S_1 - S_0$.

The Kasha rule is not valid for molecular systems with an odd number of electrons, such as aromatic radical cations. Their level pattern is different: the first doublet states, D_1 and D_0, are closely spaced whereas the $D_2 - D_1$ gap is larger. For instance, the energies of D_1 and D_2 states are 4,840 and $11,970\,cm^{-1}$ in chlorobenzene [14, 15] and 5,800 and $15,570\,cm^{-1}$ in naphthalene cations [16] so that the $D_2 \rightarrow D_0$ and $D_2 \rightarrow D_1$ emission is observed whereas that from the D_1 state is quenched.

5.2.2 Small-Molecule Properties of Large Molecules

The time evolution of large molecules is usually well described as a direct, irreversible relaxation from sparse, discrete levels $|s\rangle$ of the upper electronic state to the dense $\{m\}$ quasi-continuum of the lower state levels. The significant deviations from this scheme indicate that the efficient density of the dark levels is too low to give a continuum so that the $|s\rangle$ level is effectively coupled to a discrete set of ℓ-states.

Resonance Effects in $S \rightarrow T$ Transitions

In aromatic compounds, the energy gap between the lowest S_1 and T_1 excited $\pi\pi^*$ states is large, of the order of $10,000\,cm^{-1}$ ($\Delta E_{ST} \approx 11,500\,cm^{-1}$ in anthracene). One can thus expect that, in view of the weakness of the spin-orbit coupling between $^1\pi\pi^*$ and $^3\pi\pi^*$ states and of large energy gaps, the $S_1 \rightarrow T_1$ intersystem crossing is as slow as the previously discussed $T_1 \rightarrow S_0$ relaxation. The relaxation of the S_1 state is, however, much more rapid. In anthracene, $k_{ISC} \approx 10^7\,s^{-1}$ and its rates depend on the vibrational level of the S_1 state (below the threshold of the rapid IVR) [17, 18]. These data corroborate older reports on the temperature dependence of the anthracene triplet yield in solutions [19]. Moreover, the S-T decay rate and triplet yield Q_t is strongly changed by a substituent that has no effect on the spin-orbit coupling strength and only slightly shifts the S_1 level. The triplet yield $Q_t = 0.35$ of jet-cooled anthracene is reduced nearly to zero in 9,10 diphenyl- and 9,10 dichloro-anthracene [18].

Such a high sensitivity to small energy shifts is not expected in the case of the direct coupling of distant S_1 and T_1 states and suggests that a higher triplet state, T_i, quasi-resonant with the S_1 state is involved in a sequential $S_1 \rightarrow T_i \rightarrow T_1^{\#}$ process. The T_i levels are strongly coupled to the quasi-continuum of T_1 levels so that $T_i \rightarrow T_1$ transition is instantaneous, whereas the rate-determining step is the $S_1 \rightarrow T_i$ transition relatively rapid in view of the small $S_1 \rightarrow T_i$ energy gap and sensitive to small shifts of singlet or triplet levels, which modify the $E_{S_1} - E_{T_i}$ energy gap and induce the resonance effects.

The role of T_i states is still more important in the case of aza-aromatics with the S_1 $^1\pi\pi^*$ and T_1 $^3\pi\pi^*$ states. The rate of the $S_1 \rightarrow T_1$ transition is strongly increased when the energy of the T_2 $^3n\pi^*$ state is intermediate between those of T_1 and S_1 states. Because the $^1\pi\pi^* - {}^3n\pi^*$ coupling is stronger than the $^1\pi\pi^* - {}^3\pi\pi^*$, the intersystem crossing proceeds through the sequential $S_1(\pi\pi^*) \rightarrow T_2(n\pi^*) \rightarrow T_1(\pi\pi^*)$ channel as indicated by the El-Sayed rules [3].

There is a formal analogy between this mechanism and the previously discussed role of the overtones of C-H modes in the relaxation of the T_1 state.

The $S_2 - S_1$ Strong-Coupling Case

The deviations from the Kasha rule occur not only for the molecules with unusually large $S_2 - S_1$ gaps but also in the case of unusually small gaps. Upon the excitation of the S_2 state of naphthalene and pyrene in the gas phase [20, 21] and of naphthalene in the supersonic expansion [22], the fluorescence spectrum is composed of a strong emission from high $S_1^{\#}$ vibrational levels isoenergetic with the S_2 state and of weak bands in the spectral range where the $S_2 \rightarrow S_0$ emission (mirror image of the $S_2 \leftarrow S_0$ absorption) is expected. The decay times of both emission components are the same as upon a direct excitation of hot $S_1^{\#}$ levels. On the other hand, the 0_0^0 band of the $S_2 \leftarrow S_0$ transition in absorption is not broadened, as in the case of molecules with larger $S_2 - S_1$ gaps, but splits into a set of narrow bands [23, 24].

These specific features of the absorption and emission spectra are due the low density of $S_1^{\#}$ levels interacting with the S_2 ($v' = 0$) state. These levels do not form a quasi-continuum but a discrete manifold: the strong coupling gives rise to discrete $|\varphi\rangle = |\alpha|S_2\rangle + \sqrt{(1-\alpha^2)}|S_1\rangle$ states with $\alpha^2 \ll 1$, so that their decay rate is close to that of the lower state $k_\varphi \approx k_{S_1}$ and the intensity of the $S_1 \rightarrow S_0$ emission is predominant over the $S_2 \rightarrow S_0$ one.

Reverse $T \rightarrow S$ Relaxation and Delayed Fluorescence

The $S \rightarrow T$ transitions in large molecules are usually considered as irreversible in view of the ratio of level densities, $\rho_t/\rho_s \gg 1$. The delayed (e-type) fluorescence suggests, however, the existence of reverse $T_1 \rightarrow S_1$ transitions. In low-temperature matrices, after a pulse excitation of the S_1 state and a rapid decay of the $S_1 \rightarrow S_0$ fluorescence, the population of the T_1-state decays slowly with emission of the $T_1 \rightarrow S_0$ phosphorescence, with τ_{ph} of the order of seconds. When the sample temperature is increased so that kT is not negligibly small as compared with the $\Delta E_{T_1 S_1}$ energy gap, the phosphorescence is quenched and replaced by the long-lived delayed fluorescence with the spectrum identical with that of a normal fast fluorescence and the decay time $\tau_{df} \approx \tau_{ph}$. This phenomenon was explained [25] by the reverse $T_1^{\#} \to S_1$ intersystem crossing. The S_1 level is emptied by the fluorescence and repopulated by the $T_1 \to T_1^{\#} \to S_1$ population equilibration until the triplet reservoir is emptied. The first observation of this process was the basis of the Jablonski diagram [25].

In the more recent theories of delayed fluorescence [26], the concept of the reversible crossing is replaced by that of the mixing of quasi-resonant $T_1^{\#}$ and $S_1(0^0)$ levels. The resulting mixed states, $|\varphi\rangle = \alpha|S_1\rangle + \sum_i \beta_i |T_1(v_i)\rangle$, acquire radiative widths $\gamma^{rad} = \alpha^2 \gamma_s^{rad}$ and borrow the transition moment of the $S_1 \to S_0$ transition. This picture fits better into the theory of nonradiative transitions but the kinetic treatment of the pseudo-monomolecular process of delayed fluorescence (the energy transfer from the medium is a necessary condition) is also correct.

5.2.3 Small Molecules

We consider the coupling between individual $|vJ\rangle$ zero-order states: the bright $|v_B \approx 0, J_B\rangle |s\rangle$ level of the higher state B and a discrete manifold of dark $|\ell\rangle$ ro-vibronic levels $|v_A \gg 1, J_A\rangle$ $(J_A = J_B)$ of the lower electronic state $|A\rangle$. A significant interaction occurs only when the energy difference $\Delta E_{s\ell}$ is not much larger than the coupling matrix element, $|V_{s\ell}/\Delta E_{s\ell}| \gg 0$. The probability P that this condition is fulfilled for a pair of levels is (cf. Sect. 1):

$$P = \langle V \rangle / \langle \Delta E \rangle \approx \langle V \rangle \rho. \tag{5.16}$$

In the *weak-coupling limit.* $V\rho < 1$ and a significant two-level interaction occurs only for a fraction of the B-state levels, the other ones being practically unperturbed. In the *strong coupling case*, $V\rho \geq 1$ and each v_B, J_B level interacts with one or more v_A, J_A levels.

The Weak-Coupling Limit

The coupling of an $|s\rangle$–$|\ell\rangle$ pair gives rise to a pair of eigenstates, $|\varphi\rangle = \alpha|s\rangle + \beta|\ell\rangle$ and $|\varphi'\rangle = -\beta|s\rangle + \alpha|\ell\rangle$ (where $\alpha^2 + \beta^2 = 1$ and $\alpha^2 \geq \beta^2$), displaced from the position of unperturbed $|s\rangle$ and $|\ell\rangle$ levels. If the ℓ levels are dark and nonradiant, the intensities of the $|0\rangle \to |\varphi\rangle$ and $|0\rangle \to |\varphi'\rangle$ lines in absorption are $I_\varphi = \alpha^2 I_s \geq I_{\varphi'} = \beta^2 I_s$ and their decay rates $k_\varphi = \alpha^2 k_s \geq k_{\varphi'} = \beta^2 k_s$ are reduced with respect to that of the $|s\rangle$ level. Upon the incoherent pumping of the $0 \to \varphi'$ (extraline) and of the $0 \to \varphi$ transitions, the emission from the $|\varphi'\rangle$ level will be thus weaker and longer-lived than that of the $|\varphi\rangle$ level. The complete information about the s-ℓ coupling patterns of model molecules may be obtained by high-resolution spectroscopy and confirmed by time-resolved measurements.

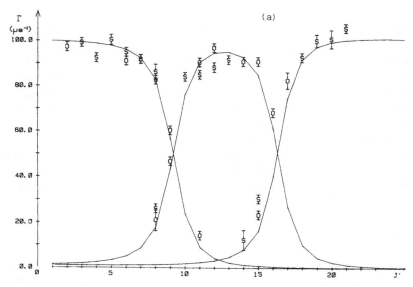

Fig. 5.4. Calculated α^2 coefficients (solid lines) and fluorescence decay rates (points) in μs^{-1} units of single rotational levels of the strongly perturbed $A^1\Pi\ v = 0$ state of CO (from [30])

Carbon Monoxide

The lowest singlet state $A^1\Pi$ of CO is strongly perturbed by the close-lying $|t_i\rangle$ levels of $a'^3\Sigma^+$, $e^3\Sigma^-$ and $d^3\Delta$ triplet states. The mixing coefficients α and β for individual $v_s = 0$–7, $J = 1$–30 rotational levels of the $A^1\Pi$ state were deduced from the high-resolution study of spectroscopic perturbations [27]. In the $v_s = 0$ state, only a few rotational levels with $J \approx 9$ and 16 are significantly perturbed, whereas for other ones, $\alpha^2 \approx 1$ and $\beta^2 \approx 0$ so that the extra lines are weak. Upon the broadband excitation of the whole set of rotational levels of the $v_s = 0$ state, the fluorescence decay is not exponential: its main component with the $\sim 10^8$ s^{-1} rate is accompanied by a weak, long tail due to the emission from weakly radiant $|\varphi', J\rangle$ levels [28, 29]. Upon a selective excitation of individual rotational levels, the decays are exponential with J-dependent rates. For most of them, $k_{\varphi,J} \approx k_s \approx 10^8$ s^{-1} and only a few levels in the vicinity of $J = 9$ and $J = 16$ show the reduced rates $k_{\varphi'} \leq k_\varphi \leq k_s$ (Fig. 5.4). The agreement between the values of α^2 and β^2 mixing coefficients deduced from level shifts (solid lines in Fig. 5.4) and observed decay rates is excellent (e.g., for $J = 9$ levels, the mixing coefficients are equal to 0.536 and 0.421, whereas their decay rates are $\sim 0.6k_s$ and $\sim 0.4k_s$) [30].

Thioformaldehyde H_2CS

This molecule is a special Creator's gift for high-resolution spectroscopists. In view of the small radiative decay rate of 7×10^3 s^{-1} [31], the homogeneous widths of rotational levels of the first excited singlet $A^1A_2\ S_1$ state ($\omega_{00} = 16,394$ cm^{-1}) state are: $\gamma_{\text{hom}} \approx 2 \times 10^{-7}$ cm^{-1}. Even the extremely weak perturbations in the $S_0 \to S_1$ spectra may be evidenced by the sub-Doppler spectroscopy [32].

The A-state is very weakly perturbed by the sparse manifolds of levels of the ground $X^1 A_1\ S_0$ state and of the close-lying $\tilde{a}\,^3 A_2\ T_1$ triplet state ($\omega_{00} = 14{,}507$ cm^{-1}). The singlet–triplet perturbations may be recognized by their sensitivity to the magnetic field (magnetic rotation) [33]. The fluorescence lifetimes of $|\varphi\rangle$ levels resulting from the perturbations are longer than those of unperturbed $|s\rangle$ levels ($\sim$140–170 μs) by factors up to 2 ($\tau \approx 200$–300 μs). The decrease of the $S_1 \to S_0$ fluorescence yield from $Q_f = 0.95$–1 to 0.6–0.7 is probably due to the collisional quenching of long-lived $|\varphi'\rangle$ levels [34].

Glyoxal CHO-CHO and Other Bicarbonyls

The case of glyoxal is more complex: beside the singlet–triplet coupling described in terms of the weak-coupling limit, the levels of the $^1 A_u\ S_1$ ($\omega_{00} = 22{,}938$ cm^{-1}) state are coupled to a quasi-continuum of predissociated ground-state levels [35]. Its methyl derivatives represent a transition from the weak-coupling to the strong-coupling case.

Glyoxal

The particularity of glyoxal is an extreme weakness of the coupling between the S_1 and the $^3 A_u\ T_1$, ($\omega_{00} = 20{,}660$ cm^{-1}) triplet state involving the second-order, spin-orbit vibronic mechanism. In spite of small $\sim 10^{-5}$ cm^{-1} homogeneous widths of levels of the long-lived S_1 state ($k_s = 4.15 \times 10^5$ and 1.04×10^5 s^{-1} for glyoxal and glyoxal-d_2, respectively [36]), no singlet–triplet perturbations could be evidenced by the high-resolution spectroscopy [37]. The fluorescence decay rate does not show any significant variation with the rotational quantum number [36].

Because $V_{\text{ST}} \ll \Delta E_{\text{ST}}$ for almost all singlet levels, the coupling could be evidenced only by the level anticrossing spectroscopy in strong magnetic fields [38] (cf. Sect. 2.2.2 and Fig. 2.4). The strict resonance ($\Delta E_{\text{ST}} = 0$) between the selectively pumped J_s singlet level and the $J_t = J_s$ triplet level is induced by the Zeeman shift of the triplet in the field B so that J_t and J_s states are mixed ($\alpha^2 = \beta^2$) and the decay rate of J_s is reduced by a factor of 2. Each anticrossing may thus be detected by measurements of the field dependence of the fluorescence lifetime in collision-free conditions.

One can determine in this way the spacing of the triplet levels $|t, J_t\rangle$, which decreases from 2 to 0.2 cm^{-1} ($\rho_t = 0.5$ to 5/cm^{-1}) with the vibrational energy of S_1 state corresponding to $E_{\text{vib}} = 2{,}276$ to 4,636 cm^{-1} in the triplet [38]. The V_{ST} values vary in width 1 to 200 MHz ($\sim 3 \times 10^{-5}$ to $\sim 10^{-2}$ cm^{-1}) limits, but most of them do not exceed 10 MHz (3×10^{-4} cm^{-1}). In view of $\langle V_{\text{ST}} \rangle \ll \langle \Delta E_{\text{ST}} \rangle$, the probability of quasi-resonances in the zero field is very low, but several such resonances have been detected and characterized by the effects of weak magnetic fields [39]. On the other hand, the presence of a few $|\varphi'\rangle$ levels with a predominant triplet character and lifetime much longer than that of $|s\rangle$ levels was evidenced by recording the excitation spectrum of the fluorescence delayed by $\Delta t \geq 10\tau_s$ with respect to the exciting pulse [40].

Bicarbonyls

The singlet–triplet coupling strength of glyoxal is not expected to be modified by methyl substitution in methyl-glyoxal (CHO-CO-CH$_3$), perdeutero-methyl-glyoxal (CHO-CO-CD$_3$), and biacetyl (di-methyl-glyoxal H$_3$C-CO-CO-CH$_3$), but the triplet level density ρ_t increases with the number of methyl groups for the same E_{vib} value. As a matter of fact, the values of the V_{ST} deduced from the Fourier transform of the decay curves of methyl-glyoxal vary randomly in the 1.3–15 MHz limits, whereas the average triplet-level densities increase with the triplet energy E_{vib} of 2,500 to 3,240 cm^{-1} from 240 to 700 in CHO-CO-CD$_3$ and from 360 to 1,200 levels per cm^{-1} in biacetyl. The decay curves recorded upon the pulse excitation of individual rotational features show a transition from the weak- to the strong-coupling case with increasing E_{vib} value [41]. The decays of the 0^0 levels of CHO-CO-CD$_3$ are either purely exponential or correspond to two-level beats, whereas for $E_{\mathrm{vib}} = 740$ cm^{-1}, all decays are modulated by two-, three- or four-level quantum beats. The complicated beat pattern of biacetyl is due to the higher t-level densities: already upon the excitation of rotational features in the 0_0^0 band, the decays are nonexponential and show many-level quantum beats [42].

The Strong-Coupling Case

When the ℓ-level density or the $V_{s\ell}$-coupling strength increases, the system evolves from the weak-coupling to the strong-coupling case ($\langle V_{s\ell} \rangle \langle \rho_\ell \rangle \geq 1$). Each $|s\rangle$ level interacts with more than one $|\ell\rangle$ level, and this interaction gives rise to a set of N $|\varphi\rangle$ states,

$$|\varphi\rangle = \alpha_{\varphi s}|s\rangle + \sum_\ell \beta_{\varphi\ell}|\ell\rangle, \tag{5.17}$$

sharing the transition moment μ_{0s} and the radiative width γ_s of the $|s\rangle$ state. If $\gamma_\ell = 0$, the average decay times, $\gamma_\varphi \approx \alpha_{\varphi s}^2 \gamma_s$, are longer than the radiative lifetime of the unperturbed $|s\rangle$ state. If it is supposed that all these states are completely mixed,

$$|\varphi\rangle = \frac{1}{\sqrt{N}}[|s\rangle + \sum_\ell |\ell\rangle], \tag{5.18}$$

we get

$$\gamma_\varphi = \gamma_s/N. \tag{5.19}$$

Such anomalously long decay times (much longer than the radiative times deduced from the intensity in the absorption spectra) were first observed for triatomic NO$_2$, SO$_2$ and CS$_2$ molecules [43]. Their specific feature is such a strong coupling between s- and sparse ℓ-levels that the $|\varphi\rangle$ states are widely spaced and incoherently excited by current light sources. The decay times observed upon selective excitation of individual rotational levels of SO$_2$ oscillate randomly between 7.5 and 68 μs [43–45], whereas the radiative lifetime of the S_1 state deduced from its absorption spectrum is of the order of 0.2 μs. The explanation of the reduced decay rates by dilution of the oscillator strength within a set of mixed states was at the origin of the theory of radiationless transitions [46].

In typical medium-sized molecules, the coupling is weaker but the level densities much larger than in triatomics and the s–ℓ interaction gives rise to a set of closely spaced absorption lines, which may be completely resolved only in the spectra of jet-cooled molecules recorded with a sub-Doppler energy resolution. In these conditions, the short-pulse excitation prepares a coherent superposition of φ-states, as evidenced by the nonexponential decay (quantum beats or biexponential decays).

We will limit our discussion to two characteristic representatives of this class of molecules: the diazines pyrazine and pyrimidine.

Pyrazine and Pyrimidine

Both compounds, very weakly fluorescent but strongly phosphorescent in rigid glasses, have been considered as belonging to the large-molecule limit until the first gas-phase experiments showed a different behaviour in the absence of environment effects. In the collision-free conditions, pyrazine [47] and pyrimidine [48] show a relatively strong fluorescence and are not phosphorescent. The fluorescence decay upon the single-vibronic-level excitation is quasi-biexponential with the lifetime of its prompt component $\tau_1 \approx 1$ ns in pyrimidine and $\tau_1 < 1$ ns in pyrazine, whereas decay time of the long-lived emission τ_2 is of the order of 100–500 ns. The spectra of both emission components are identical and correspond to the $s \to s'$ transitions. The slow fluorescence component is strongly quenched by collisions with inert collision partners (e.g., rare gases), which induce the phosphorescence. Similar behaviour was observed for biacetyl [49].

These results were explained by assuming a coherent excitation of a set of $|\varphi\rangle$ states [48]. The prompt fluorescence component was assigned to the emission from initially prepared $|s\rangle$ state—coherent superposition of $|\varphi\rangle$ states and its decay—to the loss of the initial phase agreement (dephasing). The long-lived fluorescence was assigned to an incoherent decay of $|\varphi\rangle$ states with an average $\langle \alpha^2 \rangle \, \gamma_s \approx \gamma_s/N$ rate. The efficient fluorescence quenching and phosphorescence induction by collisions was explained by a high probability of collision-induced transitions from predominantly triplet ($\alpha^2 \ll 1$) φ states to the pure triplet states.

The essential conclusions of these early works, carried out upon the unselective excitation of a number of rotational states of hot molecules, were confirmed by further studies of jet-cooled species with a narrow-band excitation of single rotational levels.

Absorption Spectra

The singlet-triplet coupling is stronger in pyrazine than in pyrimidine as shown by the fine structure of their absorption spectra.

The low rotational levels ($J \leq 3$) of the vibrationless $^1B_1$ state of pyrimidine are only weakly perturbed by the s–t coupling [50]. For instance, in the $0^0_0 R(1)$ transition, only the J, K_+, K_- (2, 2, 1) sublevel is displaced and accompanied by two extra lines

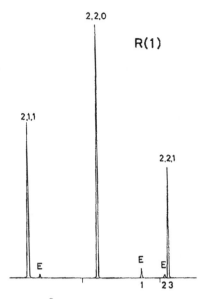

Fig. 5.5. The fine structure of the 0_0^0 $R(1)$ transition in jet-cooled pyrimidine. E indicates the extra lines (from [50])

(Fig. 5.5). An admixture of the triplet character in this state is evidenced by its lifetime of 740 ns longer than $\sim$550-ns lifetimes of unperturbed levels and by its sensitivity to magnetic field. The mixing coefficients deduced from spectral shifts of the main component of the (2, 2, 1) level cluster are

$$|\varphi_{2,2,1}\rangle = 0.945|s\rangle + 0.272|t_1\rangle + 0.185|t_2\rangle,$$

where $|t_1\rangle$ and $|t_2\rangle$ are two ro-vibronic triplet levels coupled to $|s\rangle$ by V_{ST} of 11 and 20 MHz [51]. The V_{ST} values in the 1–30 MHz limits are found for other low levels, whereas a more efficient mixing takes place at higher J values. In pyrazine, the singlet-triplet coupling is much stronger, as shown by the fine structure of rotational

Table 5.2. The s–t coupling in the pyrazine 0^0 $J = 0$ level (from [54])

Eigenstates				ZO states		
E (MHz)	I (exc)	τ (ns)	I (abs)	E (MHz)	V_{ST} (MHz)	γ (MHz)
−1,456	17	200	45	−1,286	462	5
−535	13	512	27	−502	119	1.6
−353	39	443	50	−308	105	1.6
−221	82	342	82	−98	150	2.7
−44	13	437	29	−43	Singlet	5
62	40	560	45	13	117	0.6
765	100	280	100	463	457	3
867	15	529	28	848	67	1

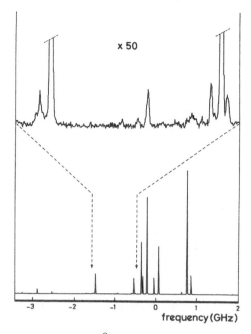

Fig. 5.6. The fine structure of the $0_0^0 P(1)$ line of jet-cooled pyrazine (from [54])

transitions in the fluorescence excitation spectra (Fig. 5.6). Already, the $P(1)$ transition to the lowest $0^0 \ J' = 0$ level is composed of 36 lines. Eight of them (the strongest ones) are listed in Table 5.2. [52]. The main parameters of the zero-order states (energies, level widths and V_{ST} coupling constants) deduced from these data using the Lawrance–Knight deconvolution procedure [53] are given in Table 5.2. V_{ST} in the 50–480 MHz range are larger by one order of magnitude than those of pyrimidine

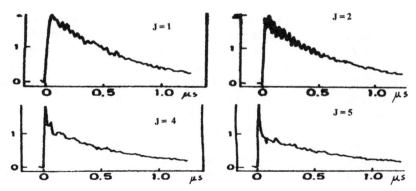

Fig. 5.7. Fluorescence decay curves from $J = 1, 2, 4$ and 5 levels of the $S_1 6a^2$ vibrational state of pyrimidine (from [55])

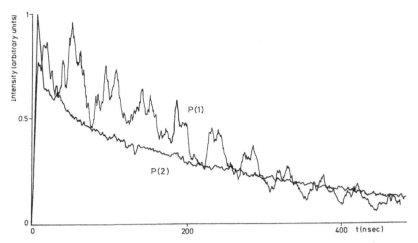

Fig. 5.8. Fluorescence decay curves from $J = 0$ and $J = 1$ levels of the $S_1 6a^1$ vibrational state of pyrazine (from [57])

[54]. The widths of the t-states, much larger than the triplet radiative widths of the order of 1 KHz, indicate an efficient $T_1 \rightarrow S_0$ nonradiative relaxation.

Decay of Excited States

The fluorescence decay times and yields depend on the vibrational energy of the excited molecule. This dependence may be easily explained by the vibrational-level density increasing with E_{vib}. More surprising is their strong dependence on the rotational quantum number.

The fluorescence decays of pyrimidine were recorded upon the coherent excitation of (J, K_+, K_-) groups of levels [55] or of single (J, K) levels [56] of the $6a^2$ state with E_{vib} of 1,225 cm^{-1} (Fig. 5.7). The decay of the $J = 0$ level is purely exponential, that of the $J = 2$ level is modulated by two-level quantum beats, whereas those of $J = 3$ and $J = 4$ are quasi-biexponential with $\tau_1 \approx 2$ ns and $\tau_2 \approx 400$ ns [54].

The fluorescence decay of pyrazine reflects a stronger s-t perturbation: the emission from the $J' = 0$ levels of the S_1 vibrational states shows many-level quantum beats, whereas the decay of the $J' = 1$ and higher rotational states is biexponential (Fig. 5.8) [57, 58]. The biexponential decays from higher vibronic levels show $\tau_1 \approx 100$ ps, significantly shorter than in the case of pyrimidine, and τ_2 of a few hundred nanosecond decreasing slowly with increasing E_{vib}.

The biexponential decays show a further J dependence for $J' \geq 5$: the yields Q_1 of the prompt emission and decay times τ_1 and τ_2 remain constant, whereas the yield of the slow fluorescence is reduced; in pyrazine, $Q_2 \approx 0.06/J'$ [59]. The same behaviour was observed in the case of pyrimidine with E_{vib} in the 1100 to 1300 cm^{-1} range [60].

The time evolution is consistent in each case with the fine structure of the absorption spectra. The quasi-exponential decay is expected in the case of the coherent excitation of the pyrimidine $J = 0$ level: as in the weak-coupling limit, the emission from φ ($\alpha^2 \approx 1$) levels with $k_\varphi \approx k_s$ is predominant whereas the long-lived emission resulting from the pumping of weak extra lines may be overlooked. In higher J states, the transition moment of the s-state is distributed among a few levels and their interference gives rise to the decay modulated by the quantum beats. Finally, this quantum beat structure is washed out in the many-level systems.

A pronounced J-dependence of the fluorescence decays may be explained by the breakdown of the $\Delta K = 0$ selection rule for the s-t coupling. Each J_s, K level is coupled with $J + 1 \, |J_t = J_s, K = 0, 1, \ldots, J\rangle$ levels. If we suppose that the s-character of the $|J_s\rangle$ state is equally distributed among $N = (J + 1) \, |\varphi\rangle$ states, the fluorescence yield Q_f and decay rate γ are

$$Q_f = \frac{\gamma_s^{\text{rad}}}{N} \Big/ \left(\frac{\gamma_s}{N} + \gamma_t \right) = \gamma_s^{\text{rad}} / (\gamma_s + N\gamma_t) \qquad \gamma_\varphi = \frac{\gamma_s}{N} + \gamma_t \qquad (5.20)$$

One can assume that for large $N \approx J$ values $\gamma_s / N \ll \gamma_t$ so that $\gamma_\varphi \approx \gamma_t$. The long fluorescence lifetime is then practically J independent, while its yield is inversely proportional to J, $Q_f \sim 1/J$.

Weak-Magnetic-Field Effects

The effect of a weak ($B \leq 200 G$) magnetic field on the fluorescence from selectively excited single J levels of pyrazine [63] and pyrimidine [55, 64] is similar to that of the increasing J value. The lifetimes τ (or τ_1 and τ_2) and the prompt emission yield, Q_1, remain unchanged, whereas Q_2 decreases monotonically and saturates at $B_{\text{max}} \approx 100\text{--}150\,\text{G}$ with $Q_f(B_{\text{max}})/Q_f(B = 0)$ ratio in the 0.1–0.3 limits for different pyrazine and pyrimidine levels. Similar effects were observed in weak ($E \leq 1\,\text{kV/cm}$) electric fields [64].

These effects are not related to the previously discussed level anticrossing in strong magnetic fields and may be explained by the decoupling of spin and rotation kinetic moments in the magnetic field (analogue of the atomic Paschen–Back effect). This decoupling modifies the selection rules for the singlet–triplet coupling and increases by a factor of 3 the number of triplet levels coupled to the singlet. The $J_t = J_s$, $m_{J_t} = m_{J_s}$ rule is attenuated to $m_{J_s} = m_{N_t} + m_{S_t}$, where m are projections of angular momenta on the field axis. Because the density of triplet levels is increased 3 times whereas the coupling constant V_{ST} is reduced by a factor of $\sqrt{3}$ (cf. Sect. 1.4.3), the decay rate, $k \sim V_{ST}^2 \rho_t$, is unchanged while the number of $|t\rangle$ levels coupled to $|J_s\rangle$: $N_{\text{eff}} = V_{ST}^2 \rho_t^2$ is increased and the fluorescence yield, $Q_f \approx \gamma_s^{\text{rad}} / N_{\text{eff}} \gamma_t$ is reduced, $Q_f(B_{\text{max}})/Q_f(B = 0) = 0.33$.

The oversimplified description of the coupling between J, K rotational states gives only a purely qualitative agreement with the experiment. A detailed theory of the coupling mechanism is needed. Application of the tier model (Sect. 3.5.2), taking into account spin orbit and Coriolis coupling, would be fruitful.

5.3 The Environmental Effects on Electronic Relaxation

5.3.1 Large and Intermediate-Sized Molecules

The necessary condition of an irreversible electronic relaxation is the coupling of the bright state $|s\rangle$ to the set of lower state levels, which behave as a dissipative continuum $\{m\}$. This condition is fulfilled in an isolated large molecule so that the relaxation mechanism (its rate) is not significantly modified by medium-induced vibrational relaxation and broadening of the m levels.

In contrast, these level-broadening effects play an essential role in the case of intermediate-sized molecules described in terms of the strong-coupling case. In pyrazine or pyrimidine, the $|s\rangle$ ro-vibronic levels of the S_1 state are coupled to a few $|\ell\rangle$ levels, which constitute a small fraction of a dense set of triplet levels. The collisions with inert partners induce the vibrational and rotational relaxation within the triplet manifold coupling the $|\ell\rangle$ levels to the continuum of kinetic energy. This process may be described as a sequential $s \rightarrow \ell \rightarrow \{m\}$ relaxation and is accounted for by broadening of $|\ell\rangle$ levels. At a sufficiently high gas pressure, the $\gamma_\ell \geq \langle \Delta E_{\ell\ell'} \rangle$ limit is attained so that the ℓ levels form a continuum. For typical intermediate-sized molecules, the transition from collision-free conditions to condensed phases corresponds to the transition from strong-coupling case to the large-molecule limit.

5.3.2 Small Molecules

Intrinsic and Collisional Coupling

The absence of the electronic relaxation in isolated small molecules is due not only to the small value of the $V_{s\ell}/\Delta E_{s\ell}$ parameter but also, very often, to the weakness of the $V_{s\ell}$ coupling. The answer to the question whether the s-ℓ coupling may be induced by a collision is different for transitions within a single spin manifold (the collision-induced internal conversion CI IC) and between different spin manifolds (the collision-induced intersystem crossing CI ISC).

Collision-Induced Internal Conversion

The electronic states differing by their orbital symmetry are uncoupled in a rigid, nonrotating molecule. The coupling is induced in an isolated polyatomic molecule by nontotally symmetric vibrations, but this mechanism is not operating in a diatomic molecule because the only vibrational mode is totally symmetric. The coupling $\langle s|V^{\text{coll}}|\ell\rangle \neq 0$ is, however, induced when the symmetry of the X_2 molecule is lowered in a $X_2.C$ complex. In a T-shaped (C_{2v}) complex, the Σ_g and Σ_u states become A_1 and B_2 and may be mixed by the B_2 bending vibration, whereas in a linear configuration of $C_{\infty v}$ symmetry, they belong to the same Σ species. The internal conversion is thus allowed even in the absence of the intrinsic coupling ($\langle s|V^0|\ell\rangle = 0$).

Experience shows that the efficiency of CI IC is practically independent of the s-ℓ coupling strength in the isolated molecule. The cross-section of the CN $A^2\Pi \rightarrow X^2\Sigma^+$ CI IC induced by the Ar atom is nearly the same for strongly perturbed and unperturbed rotational levels of the $v = 3$, $v = 7$ and $v = 8$ states [65]. Moreover, the efficiency of the collision-induced $A^2\Pi_u \rightarrow X^2\Sigma_g^+$ relaxation in the centrosymmetric N_2^+ in which the g-u coupling is strictly forbidden in collision-free conditions is the same as that of the $A^2\Pi \rightarrow X^2\Sigma^+$ in the isoelectronic CN molecule [66].

Collision-Induced Intersystem Crossing

As long as the electronic wave functions may be represented as products of space coordinate and spin functions, the $\Delta S = 0$ Kronig rule is valid and external perturbations cannot induce the coupling between pure spin (e.g., singlet and triplet) states,

$$\langle s | V^{\text{coll}} | t \rangle = 0,$$

but only between mixed states: $|\varphi\rangle = \alpha|s\rangle + \beta|t\rangle$ and $|\varphi'\rangle = \alpha'|s\rangle + \beta'|t\rangle$ states,

$$\langle \varphi | V^{\text{coll}} | \varphi' \rangle = \alpha\alpha' \langle s | V^{\text{coll}} | s' \rangle + \beta\beta' \langle t | V^{\text{coll}} | t' \rangle. \tag{5.21}$$

The cross-sections of a collision-induced transition from the $|\varphi\rangle$ state to pure triplet ($\alpha' = 0$) and pure singlet ($\beta' = 0$) levels are

$$\sigma_{\varphi \rightarrow s} = \alpha^2 \sigma_{\text{rot}}, \qquad \sigma_{\varphi \rightarrow t} = \beta^2 \sigma_{\text{rot}}, \tag{5.22}$$

where σ_{rot} is the cross-section for relaxation (essentially rotational) within the same spin manifold, supposed to be the same in the singlet and triplet manifolds [68].

The essentially singlet s' state ($\beta^2 \ll \alpha^2$) is thus relaxed to one of the pure singlet levels with probability much higher than that of the $s' \rightarrow t$ relaxation ($\sigma_{\text{ISC}} \ll \sigma_{\text{rot}}$), whereas an essentially triplet t' ($\beta^2 \gg \alpha^2$) level relaxes to the t-manifold.

The efficient quenching of the long-lived fluorescence of SO_2 [69], pyrazine [47]] and pyrimidine [70] by collisions with inert partners is due to an essentially triplet character of φ states resulting from the mixing of a single $|s\rangle$ state with a large number of $|\ell\rangle$ states. The $\varphi \rightarrow t$ relaxation to the dense set of nonfluorescent triplet levels is much more probable than transitions between sparse $|\varphi\rangle$ levels.

The validity of Eq. 5.22 may be checked by determining the correlation between average values of $\langle\sigma_{\text{ISC}}\rangle$ and $\langle\beta^2\rangle$ for individual vibronic levels of a molecule, provided that the ISC rate is larger than that of the vibrational relaxation and small as compared with the rotational relaxation rate ($\sigma_{\text{rot}} \gg \sigma_{\text{ISC}} \gg \sigma_{\text{vib}}$. The $\sigma_{\text{ISC}}(v)$ values for the $v = 0$–7 vibrational levels of CO $A^1\Pi$ state determined by the fluorescence quenching in CO + He collisions [71] and $\langle\beta_v^2\rangle$ average values of mixing coefficients deduced from spectroscopic data [27] are given in Table 5.3. The average values of $\langle\sigma_{\text{ISC}}\rangle$ are correlated with $\langle\beta^2\rangle$ but the proportionality relation, $\sigma_{\text{ISC}} \sim \beta^2$ suggested by Eq. 5.22, is not found. The β^2 coefficients vary in wide limits for rotational levels of the same vibronic state and most of the s' levels are pure singlet ($\beta^2 \approx 0$) levels and only a few of them are significantly mixed (cf. Sect. 5.2.2 and Fig. 5.4). For these levels, $\sigma_{\varphi \rightarrow s} \approx \sigma_{\varphi \rightarrow t}$, so that they play a role of gates between the singlet and triplet manifolds. One can easily show that the efficiency of a process involving two subsequent collisional transitions $s' \rightarrow gate$ and $gate \rightarrow t'$ is much larger than that of a direct $s' \rightarrow t'$ transition

Table 5.3. Averaged mixing coefficients β^2 and σ_{ISC} cross-sections in CO-He collisions for CO $^1\Pi$ state

v	Perturbing level	$\langle \beta_v^2 \rangle$	$\sigma_{ISC}(A^2)$
0	$e^3\Sigma^-$ $(v = 1)$	0.102	3.4
	$d^3\Delta$ $(v = 4)$		
1	$d^3\Delta$ $(v = 5)$	0.062	2.7
2	$e^3\Sigma^-$ $(v = 4)$	0.0044	0.9
3	$d^3\Delta$ $(v = 8)$	0.003	0.45
4	$a'^3\Sigma^+$ $(v = 14)$	0.012	1.35
5	$e^3\Sigma^-$ $(v = 8)$	0.002	0.55
6	$d^3\Delta$ $(v = 12)$	0.102	3.6
	$a'^3\Sigma^+$ $(v = 17)$		
7	$e^3\Sigma^-$ $(v = 11)$	0.001	0.5

In the glyoxal ^1A_u state, $\sigma_{ISC}/\sigma_{rot} = 0.04$ is deduced from ISC measurements in a complete contradiction, with $\langle \beta^2 \rangle \approx 10^{-5}$ from the level anti-crossing studies [39]. Intersystem crossing proceeds through the gate states as evidenced by the enhanced efficiency of CI ISC upon excitation of single rotational levels in the single-collision conditions [72].

5.3.3 Pathways of Electronic Relaxation

Reversibility of s-ℓ Transitions

In light diatomic molecules with large $E_{v,v-1}$ energy gaps, the vibrational relaxation is usually slow as compared with the s-ℓ electronic relaxation ($k_{vib} \ll k_{ISC}$). A pseudo-equilibrium between populations of quasi-resonant vibronic states, $|s, v_s\rangle$ and $|\ell, v_\ell\rangle$, may thus be established during the lifetime of the vibronic level,

$$\frac{N_\ell}{N_s} = K = \frac{k_{s\to\ell}}{k_{\ell\to s}} = \frac{\rho_\ell}{\rho_s}e^{-(E_\ell - E_s)/kT}. \tag{5.23}$$

When $\rho_\ell \approx \rho_s$ and the energy gap $|E_\ell - E_s|$ is not very large as compared with kT, the population N_s and N_ℓ are of the same order of magnitude.

Such a quasi-equilibrium between populations of $a^3\Pi_u$ and $X^1\Sigma_g^+$ states of C_2 with $K \approx 3$ for the C_2-O_2 collisions was deduced from reactivity of C_2 [73]. A true equilibrium between closely spaced $^1A_1$ and $^3B_1$ states of methylene CH_2 was evidenced: singlet or triplet methylene is initially prepared, but at a sufficiently high pressure of the inert collision partner (SF_6), the singlet-to-triplet concentration ratio is the same [74]. The reversibility of transitions between the $A^1\Pi$ $v = 0$ and $e^3\Sigma^-$ levels of CO induced by collisions with rare-gas atoms is directly evidenced by the pressure dependence of the decays of the populations of the s' and t' levels. The decay curves may be fitted in a wide range of quencher pressures by kinetic equations involving rate constants of the radiative decay of the s-state k_s^{rad}, the $k_{s\to t}$ and $k_{t\to s}$

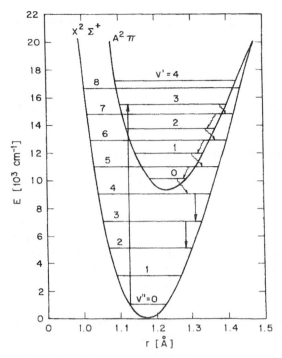

Fig. 5.9. Potential-energy curves of CN X and A states and schemes of the relaxation cascades (from [77])

intersystem crossing rates and the rates k_t^{vib}, $k_s^{\text{vib}} = 0$ of the vibrational relaxation in the singlet and triplet manifolds [71].

Vibrational Relaxation Via Reversible s-ℓ Transitions

The reversibility of $s \leftrightarrow \ell$ transitions opens an indirect channel of vibrational relaxation, $|s, v\rangle \rightarrow |\ell\rangle \rightarrow |s, v - 1\rangle$, which may be more rapid than the direct $|s, v\rangle \rightarrow |s, v - 1\rangle$ relaxation in molecules with large $E_v - E_{v-1}$ gaps. So, upon the excitation of the $A^1\Pi$ $v = 1$ level of CO in the CO + Ar gaseous mixture, the relaxed fluorescence of the $v = 0$ level builds up with a large delay with respect to the decay of the $v = 1$ emission, which is incompatible with a direct $v = 1 \rightarrow v = 0$ relaxation path. Such a delay may be explained by a sequence of collision-induced intersystem transitions: from the $A^1\Pi$ ($v = 1$) to a long-lived triplet level and back to the $A^1\Pi$ ($v = 0$) state [29].

The same mechanism governs the vibrational relaxation of CN in Ne.CN complexes [75, 76] and rare-gas matrices [77] (Fig. 5.9). In matrices, the direct relaxation in the ground $X^2\Sigma^+$ electronic state is slow, as shown by millisecond lifetimes of $v = 4$ and $v = 3$ levels, whereas those of higher levels are shorter by three orders of magnitude (Table 5.4). The indirect relaxation mechanism was evidenced by measurements of the time dependence of populations of $A^2\Pi v \leq 3$ and $X^2\Sigma^+ v \leq 8$

levels upon the initial excitation of the $A^2\Pi$ $v = 4$. The rates of individual steps of the $A^2\Pi(v) \rightarrow X^2\Sigma^+(v') \rightarrow A^2\Pi(v-1)$ chain are related to the $E_v - E_{v'}$ and $E_{v'} - E_v$ energy gaps by the formula

$$k_{\mathrm{rel}} = A|\langle v|v'\rangle|^2 e^{-\beta\Delta E}, \tag{5.24}$$

taking into account the Franck–Condon factor between initial and final state.

5.4 Intersecting Potential-Energy Surfaces

The intersection of diabatic surfaces, U_a and U_b—the avoided intersection of adiabatic surfaces U_A and U_B, gives rise to two important photophysical processes:

- the *electronic predissociation* when the lower A state is dissociative or weakly bound so that the energy of v_B level exceeds the bonding energy of the A state,
- the *ultrafast $B \rightarrow A$ relaxation* when the high vibrational levels of the A-state are bound but behave as a dissipative continuum. This process thus takes place only in polyatomic molecules with relatively large density of v_A levels isoenergetic with the v_B one.

The mechanism of predissociation is essentially the same in diatomic and poly-atomic molecules, but the time evolution of polyatomic systems is much more complex because of the interference between different relaxation channels (IVR, electronic relaxation, electronic and vibrational predissociation, ...). We begin, therefore, by discussion of the electronic predissociation—the unique nonradiative decay channel of a diatomic molecule in terms of one-dimensional potential curves. The main conclusions may then be extrapolated to the surface crossing in polyatomic systems.

5.4.1 Curve Crossing and Predissociation of Diatomics

The electronic predissociation is energetically allowed when the electronic energy of the higher (b) electronic state exceeds the dissociation onset of the lower a state. The bound ro-vibronic v_b level of the XY^* molecule is then isoenergetic with the $\chi_a(E)$ state of the continuous spectrum of the $X + Y$ pair (where X and Y are atoms or

Table 5.4. Vibrational levels, energy gaps and lifetimes of excited CN in Ne matrix (from [77])

$v A^2\Pi$		$v' X^2\Sigma^+$	ΔE cm^{-1}	τ (μs)
4	$\rightarrow$	8	542	$\sim$0.01
3	$\leftarrow$	8	1,168	7.5
3	$\rightarrow$	7	692	0.09
2	$\leftarrow$	7	1,043	5.0
2	$\rightarrow$	6	843	0.67
		4	1,964	2,800
		3	1,990	5,400

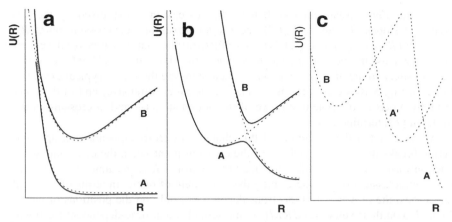

Fig. 5.10. The direct predissociation in the glancing (a) and peaked (b) case of the curve intersection and the indirect predissociation (c)

atom groups). The rate of an irreversible transition from the discrete v_b state to the a continuum (i.e., of the $XY^* \rightarrow X + Y$ predissociation) is, according to the Fermi golden rule, proportional to the square of the coupling matrix element,

$$V^2 = |\langle \chi_{v_b}|V_{ab}|\chi_a(E)\rangle|^2, \qquad (5.25)$$

where V_{ab} is the diabatic electronic coupling matrix element similar to the one in Eq. (5.12) and $|\chi_{v_b}\rangle$ and $|\chi_a(E)\rangle$ are nuclear functions. When the strength of the electronic coupling varies slowly with the R_{XY} nuclear coordinate, Eq. (5.25) can be rewritten as

$$V^2 = V_{ab}^2 |\langle \chi_{v_b}|\chi_a(E)\rangle|^2. \qquad (5.26)$$

The overlap between localized wave function of the bound $|v_b\rangle$ state and rapidly oscillating function $\chi_a(E)$ is close to zero (cf. Sect. 1.6.2. and Fig. 1.8) except for the case of the intersection of U_b and U_a potential surfaces, so that the turning points of χ_{v_b} and of $\chi_a(E)$ coincide. If the crossing occurs at the energy E_c higher than the dissociation limit of the a state, the coupling of the discrete v_b level with the continuum of the a state gives rise to the *direct predissociation* of a set of b-state levels with energies $E \approx E_c$.

The *indirect (or accidental) predissociation* [78] occurs when the $U_b(R)$ curve crosses that of an intermediate a' state that is bound but predissociated because of the crossing of the $U_{a'}(R)$ curve with that of the dissociative state a. The predissociation is thus induced by the *sequential coupling* of the bound v_b level to an intermediate $v_{a'}$ discrete level coupled to the dissociative continuum. The typical case is that of a low vibronic level of the S_1 state interacting with the high ground-state $S_0^{\#}$ level broadened by the vibrational predissociation.

The third factor determining the probability of the $b \rightarrow a$ transition is the extension of the R range, in which the coupling between the a and b states V_{ab} is nonnegligible as compared with $\Delta U_{ab} = |U_a(R) - U_b(R)|$ energy gap. In the Landau–Zener treatment (see below), this extension is expressed by the difference of slopes of diabatic $U_a(R)$ and $U_b(R)$ curves in their crossing point R_c: $|dU_a(R)/dR - dU_b(R)/dR|_{R_c}$. It is thus useful to differentiate two types of the curve-crossing events: the *peaked* intersection, where the signs of the dU/dR derivative are opposite, and the *glancing* one, where their signs are the same. Typical examples shown in Fig. 5.10 are the crossing events in the inner (repulsive) and in the outer (attractive) part of the potential curve of the bound state ($-$ and $+$ crossing in the Mulliken [79] terminology).

As indicated at the beginning of this chapter, the Born–Oppenheimer approximation breaks down in the vicinity of the crossing point when the electronic wave function varies so rapidly with R that the $(1/\mu_{XY})\partial\Psi/\partial R$ term cannot be neglected. In the semiclassical picture describing the movement of nuclei in terms of classical trajectories, the movement of electrons depends not only on the positions of nuclei R but also on their velocities dR/dt. In the semiclassical time-dependent treatment based on pioneering Landau [80] and Zener [81] studies of inelastic collisions, the system going through the crossing point may follow either the diabatic or the adiabatic curve with probability of a jump between U_a and U_b surfaces given by the formula

$$P_{ab} = \exp\left[-\frac{4\pi^2}{hv_c(E,J)} \frac{V_{ab}^2}{|dU_a/dR - dU_b/dR|_{R_c}}\right], \tag{5.27}$$

where v_c is the velocity in the crossing point. This model was developed in a series of further works (cf. [82–86]). Equivalent results may be obtained in an all-quantum treatment involving previously introduced (Sect. 5.1) coupled equations.

5.4.2 Coupled Equations Near an Intersection

To provide some insight into the dynamics of a curve crossing, we consider a simple model that can be solved analytically. The system of coupled equations to be solved in a diabatic representation is of the form (cf. Eqs. (5.4) and (5.8))

$$\chi_a'' + (E - U_a)\chi_a = U_{ab}\chi_b$$
$$\chi_b'' + (E - U_b)\chi_b = U_{ab}\chi_a, \tag{5.28}$$

where E and U are [energy]$/(\hbar^2/2\mu)$. We assume three distinct zones, (i) $0 \leq R \leq R_1$ (left of the intersection point), (ii) $R_1 \leq R \leq R_2$ (interaction region) and (iii) $R_2 \leq R \leq \infty$ (right of the intersection point). We further assume the electronic potentials $U_a(R)$ and $U_b(R)$ to be constant in zones 1 and 2 ($U_a(R) \equiv U_a$, $U_b(R) \equiv U_b$) and the coupling to be zero, $U_{ab} \equiv V_{ab} = 0$, so that, for given (positive) total energy E, the nuclear motion will correspond to that of a free particle with constant wave number $k_a = \sqrt{E - U_a}$ and $k_b = \sqrt{E - U_b}$, associated with one or the other potential, respectively. In the interaction zone 2, we take $U_{ab} = \text{const} \neq 0$ and we set $U_a = U_a = U$ in this region, with $k = \sqrt{E - U}$.

It is easy to see that the solutions of the coupled equations in the three zones are

Zone 1 $\chi_a^{(1)}(R) = c_a^{(1)} \sin k_a R$

$$\chi_b^{(1)}(R) = c_b^{(1)} \sin k_b R$$

Zone 3
$$\chi_a^{(3)}(R) = c_a^{(3)} \sin (k_a R + \delta_a)$$

$$\chi_b^{(3)}(R) = c_b^{(3)} \sin (k_b R + \delta_b)$$

Zone 2
$$\chi_a^{(2)}(R) = \frac{1}{2}\left[c^- \sin(k^- R + \delta^-) + c^+ \sin(k^+ R + \delta^+) \right]$$

$$\chi_b^{(2)}(R) = \frac{1}{2}\left[c^- \sin(k^- R + \delta^-) - c^+ \sin(k^+ R + \delta^+) \right], \quad (5.29)$$

where $k^- = \sqrt{k^2 - U_{ab}}$ and $k^+ = \sqrt{k^2 + U_{ab}}$. The coefficients $c_i^{(j)}$, c^-, c^+ and the phase shifts δ^-, δ^+, δ_a and δ_b are arbitrary at this point and must be determined by matching the solutions at the points R_1 and R_2. δ_a and δ_b in particular are the scattering phase shifts resulting from the interaction in zone 2.

The solutions in zone 2 correspond to a periodic oscillation of the amplitude between the channels 1 and 2. Indeed, when $c^- \approx c^+$, we have

$$\chi_a^{(2)}(R) \approx \sin\left[\frac{k^- + k^+}{2} R + \cdots \right] \cos\left[\frac{k^- - k^+}{2} R + \cdots \right]$$

$$\chi_b^{(2)}(R) \approx \cos\left[\frac{k^- + k^+}{2} R + \cdots \right] \sin\left[\frac{k^- - k^+}{2} R + \cdots \right], \quad (5.30)$$

i.e., the amplitude is transferred forth and back as R increases between the two channels, with an oscillation frequency corresponding to $(k^- - k^+)/2$. When, on the other hand, $c^- \neq c^+$, the oscillations are reduced, and when $|c^-| \gg |c^+|$ or $|c^-| \ll |c^+|$, the system becomes uncoupled.

In order to determine c^- and c^+, we choose $c_a^{(1)} = 1$ $c_b^{(1)} = 0$ in zone 1, i.e., we assume the system to be entirely in channel 1 left of the crossing point (before the passage through the interaction region). At the boundary $1 \leftrightarrow 2$, we must therefore have $\chi_b^{(2)}(R_1) = \chi_b^{(2)\prime}(R_1) = 0$ (where the prime refers to differentiation with respect to R), which implies that

$$c^- \sin(k^- R + \delta^-) = c^+ \sin(k^+ R + \delta^+)$$

$$c^- k^- \cos(k^- R + \delta^-) = c^+ k^+ \cos(k^+ R + \delta^+) \quad (5.31)$$

for $R = R_1$. The solutions $\chi_a^{(1)}(R_1)$ and $\chi_a^{(2)}(R_1)$ must also be matched by requiring their logarithmic derivatives to be equal at $R = R_1$, i.e., $\chi_a^{(1)\prime}(R_1)/\chi_a^{(1)}(R_1) = \chi_a^{(2)\prime}(R_1)/\chi_a^{(2)}(R_1)$. This condition leads us to

$$\delta^+ = (\bar{k}^+ - k^+)R_1, \quad \text{where } \tan\left[\bar{k}^+ R_1 \right] \equiv \frac{k^+}{k_a} \tan\left[k_a R_1 \right]$$

$$\delta^- = (\bar{k}^- - k^-)R_1, \quad \text{where } \tan\left[\bar{k}^- R_1 \right] \equiv \frac{k^-}{k_a} \tan\left[k_a R_1 \right]. \quad (5.32)$$

We substitute this result into Eq. 5.31 and find that

$$\frac{c^-}{c^+} = \frac{\sin\left[\bar{k}^+ R_1\right]}{\sin\left[\bar{k}^- R_1\right]}. \tag{5.33}$$

Equation 5.33 tells us, qualitatively, that when the coupling $|U_{ab}|$ is *small*, we have $k^+ \approx k^-$, $\bar{k}^+ \approx \bar{k}^-$ and hence $c^- \approx c^+$, and therefore there will be a periodic transfer of amplitude from channel 1 to channel 2 and back as R increases. On the other hand, when $|U_{ab}|$ is *large*, there will be no transfer. This observation is in agreement with the generally known fact that, when curve or surface crossings are strongly avoided, there will be no, or only weak, coupling of the vibrational motions in the two electronic states.

Consider now the situation of a weakly avoided crossing, i.e., small $|U_{ab}|$ (Fig. 5.2 (b)). In that event, as we have already seen, Eq. 5.30 applies and a transfer of amplitude between the two channels takes place. The net result of the transfer, which is periodic, depends on the range over which coupling occurs, i.e., the difference $R_2 - R_1$. Equation 5.29 indicates that, when the interaction range is small enough so that we can replace the sin function by its argument, the amplitude transferred to channel 2 ($\chi_b^{(2)}$) will be of the order of $(k^- - k^+)(R_2 - R_1)/2$. Expanding $k^- = \sqrt{k^2 - U_{ab}}$ and $k^+ = \sqrt{k^2 + U_{ab}}$ to first order in U_{ab} and squaring, we obtain the probability of transfer in this limit as

$$P_{a \to b} \approx \frac{U_{ab}^2}{4k^2}(R_2 - R_1)^2. \tag{5.34}$$

This result can also be obtained by introducing the particle velocity, $v_c = \hbar k/m$, and applying time-dependent perturbation theory with a constant perturbation active during the passage time between R_1 and R_2. Equation 5.34 is reminiscent of the Landau–Zener formula in its original version (Eq. 5.27): the probability of transfer is proportional to the square of the interaction. It is the *smaller* the *larger* is the particle momentum $\hbar k$, and it is the *larger* the *larger* is the interaction region. In the Landau–Zener treatment, the extent of the interaction region, analogous to $R_2 - R_1$, is given by the inverse of the difference of the slopes of the two interacting potentials, $1/|dU_a/dR - dU_b/dR|$. In other words, a glancing intersection translates into a large value of $R_1 - R_2$ in the present model, while a peaked intersection corresponds to a small value of this quantity.

We considered here a stationary state of the system, but this treatment may be easily extended to a coherent superposition of states forming a vibrational wave packet. In the interaction zone, the wave packet splits into two packets moving on two different surfaces.

5.4.3 Experimental Studies of Predissociation

We have at our disposal a large amount of data on predissociation of diatomic molecules deduced either from lifetime measurements or from the line widths in absorption spectra.

A slow predissociation with the rate k^d of the same order of magnitude as the radiative rate k^{rad} of the unperturbed electronic state may be evidenced by the reduced yield $Q_f = k^{rad}/(k^{rad} + k^d)$ and increased decay rate $k^f = (k^{rad} + k^d)$ of the fluorescence from predissociated levels compared with $Q_f = 1$ and $k^f = k^{rad}$ of unperturbed ones. When $k^d \gg k^{rad}$, the fluorescence is entirely quenched but k^d values may be deduced from homogeneous widths of absorption lines, $\gamma = \hbar k_d$, as long as the rotational structures may be resolved. This structure is washed out for $\gamma > 2BJ$, i.e., when the level lifetime $\tau^f \approx 1/k^d$ is shorter than the classical rotational period. In the same way, the washing out of the vibrational structure indicates that the dissociation occurs during one vibrational period. In this case, the widths of individual ro-vibronic levels strongly overlap and a selective excitation of a single state is not possible; even the narrow-band light source prepares a coherent superposition of states forming a *vibrational wave packet*. Its time evolution may be considered in the diabatic basis as an ultrafast predissociation or in the adiabatic basis as a direct dissociation of the states with energies close to the top of an energy barrier.

Direct Predissociation

A good example of the weak direct predissociation is that of the $B^3\Pi(0_u^+)$ state of iodine I_2, due to its perturbation by the dissociative state $^1\Pi(1u)$, forbidden in the first approximation and allowed by second-order effects [87]. The fluorescence decay time depends on the vibrational level [88, 89] (Fig. 5.11) with two pronounced maxima of k^d: a narrow one at $v = 5$ and a broad one at $v \approx 25$, which suggest a double crossing of the diabatic potential curves. The first of them seems to correspond to the peaking and the latter one to the glancing crossing case represented in the inset of Fig. 5.11.

The major part of investigated cases of the intermediate-strength and strong perturbation cases (cf. [90]) correspond to the peaked crossing in the attractive part of the potential curve of the bound state (the + case).

Indirect (Accidental) Predissociation

The direct coupling of a bound electronic state to the dissociative continuum $\{m\}$ perturbs a whole set of ro-vibronic s levels in the vicinity of the curve crossing. A sudden broadening of one or a few isolated levels of the bound state is unexpected. Such a behaviour was explained by a *sequential coupling* of individual s levels to close-lying levels ℓ of another electronic state that is bound but coupled to the continuum $\{m\}$. The $|s\rangle \leftrightarrow |\ell\rangle \leftrightarrow m$ coupling is efficient only for a small fraction of s levels that are accidentally in quasi-resonance with ℓ levels. For instance, the $v = 3$ level of the $b^1\Pi_u$ state of N_2 is weakly coupled by the spin-orbit interaction to the $C^3\Pi_u$ level predissociated by the $C'^3\Pi_u$ continuum [90].

A detailed study of the indirect predissociation by the level-anticrossing technique was reported for formaldehyde H_2CO and formaldehyde-d_2 D_2CO. The ro-vibrational levels of the S_1 $^1A_2$ state (s levels) are isoenergetic with high ground-state levels $S_0^{\#}$ (ℓ levels), which are predissociated by the dissociative continuum $\{m\}$ of H_2+ CO fragments separated by an energy barrier [91]. The $S_0^{\#}$ levels are so widely spaced

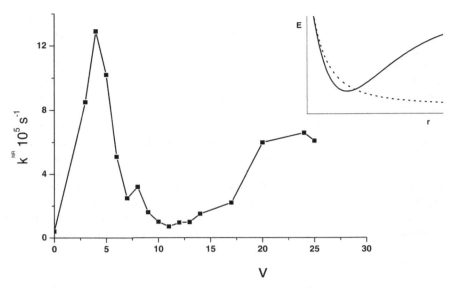

Fig. 5.11. The v dependence of the predissociation rate of the $B^3\Pi(0_u^+)$ state of iodine. Inset: the artist view of the curve crossing

that the $S_1 - S_0^\#$ perturbations of (vJK) levels of the S_1 state correspond to the weak-coupling limit with the $V_{s\ell}/\Delta E_{s\ell}$ parameter strongly varying from state to state. On the other hand, the ℓ-$\{m\}$ coupling implies an indirect $s \to \ell \to \{m\}$ predissociation of the S_1 levels. An unusually large random variation of decay rates of individual J, K levels of the 0^0 state: between 2.8×10^5 and 1.2×10^7 s^{-1} for H_2CO reduced in D_2CO to 1.2–1.8×10^5 s^{-1} [92, 93] was tentatively explained by such a coupling scheme. This assumption was confirmed by the level anticrossing in a strong electric field. Because the permanent dipole moments in S_0 and S_1 states μ_0 and μ_1 are different, the electric field $\mathcal{E}$ induces a relative shift of the S_1 and S_0 level sets. The slowly varying field brings the selectively excited $|s, v, J\rangle$ level to the resonance with different $|\ell, v', J' = J\rangle$ levels. Each resonance is indicated by a decrease of the fluorescence intensity and lifetime [93], the width of the signal being a measure of the coupling strength. As seen in Fig. 5.12, the broad anticrossing lines of the H_2CO spectrum overlap, but the spectrum of D_2CO is well resolved and the values of the coupling constants, $V_{s\ell}$, and of level widths, γ_ℓ, were measured for a whole set of S_0 ℓ-levels. Both parameters vary in very wide limits, showing that the high $S_0^\#$ levels conserve their individual properties in spite of the anharmonic and Coriolis mixing. The deuterium effect is explained by the mass dependence of the tunneling rate through the energy barrier.

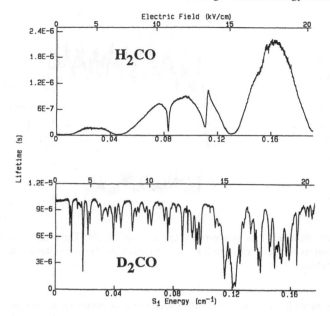

Fig. 5.12. Level anticrossing spectra of H_2CO and D_2CO in the electric field (from [93])

Ultrafast Predissociation

Time-Resolved Pump-Probe Spectroscopy at Femtosecond Scale

While the slow predissociation is described by assuming a selective excitation of a single ro-vibronic level, the short (subpicosecond) pulse excitation prepares a coherent superposition, $\Psi(0) = \mu F|0\rangle$, of molecular states $|n\rangle$ contained in the energy band $\delta\omega_{exc}$ with amplitudes proportional to μ_{0n}. The shape of the energy band, $F_0(\omega)$, is given by convolution of the exciting pulse with the vibrational wave function of the ground-state $|0\rangle$ level. The initially prepared state is the superposition $\Psi(0) = \sum_n F_0(\omega_{0n})\mu_{0n}|n\rangle$ of free and bound vibrational levels of the excited A^* electronic state—a vibrational wave packet $\Psi(Q, t)$ moving at the adiabatic potential surface A^*. Its displacement is monitored by the probe pulse with a variable time delay Δt_{probe} and frequency ω_{probe}, inducing a transition from the A^* to a higher A^{**} state. If this frequency corresponds to the energy difference between the potential-energy curves of the A^* and higher A^{**} states in the point Q_i:

$$\omega_{probe} = U_{A^{**}}(Q_i) - U_{A^*}(Q_i),$$

the number of molecules transferred to the A^{**} state (deduced from the intensity of the $A^{**} \to 0$ fluorescence or from ion signal resulting from the $A^{**} \to A^+ + e^-$ or $A^{**} + \hbar\omega \to A^+ + e^-$ ionization) is proportional to the square of the amplitude, $|\Psi(Q_i, t_{probe})|^2$, of the wave packet. By varying ω_{probe} and Δt values, one can follow the migration of the packet on the $U_{A^*}(Q)$ surface. This technique may be applied to

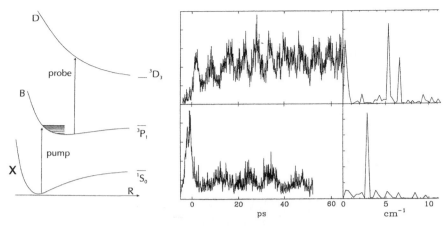

Fig. 5.13. (Left) Scheme of the pump-probe experiment. (Right) The time dependence of the probe signal and its Fourier transform for low (top) and high (bottom) vibronic levels of the B state of HgAr (from [95])

bound states as well as to the direct dissociation and predissociation processes. We will briefly discuss two former cases as introduction to the predissociation case.

Bound States of a Diatomic Molecule

In the strictly harmonic potential, the $\Delta E_{v,v-1} = \hbar\omega$ level spacing is constant and the initially prepared wave packet oscillates between the inner and outer turning points, r_i and r_o, of the $U(r)$ potential curve with frequency ω and unchanged initial width. In an anharmonic potential, the level spacing varies with v, $\omega_v \approx \omega_e[1 - x(2v+1)]$, which implies a dephasing of oscillations so that the widths of recurrences increase with time and the depth of modulation is reduced. Because ω_v is smaller and the $\Delta\omega_{v,v-1}/\omega_v$ ratio larger for large v, the oscillation period is longer for higher vibrational levels and their dephasing is more rapid. Such a behaviour was observed for the $B^3\Pi(0_u^+)$ state of I_2 [94] and for the A state of HgAr complex correlated to the $^3P_1$ state of Hg [95]. In HgAr, the oscillation period is increased from $\sim$6 ps upon excitation of low vibrational levels to $\sim$18 ps upon excitation just below the dissociation onset; this difference corresponds to the level spacing reduced from $\sim$1 cm^{-1} to $\sim$0.3 cm^{-1} (Fig. 5.13).

Direct Dissociation

The wave packet prepared by optical excitation of a dissociative state corresponds to a coherent superposition of the states of the continuum and its time evolution is aperiodic.

A good example is ICN [96], which may be considered as a quasi-diatomic I-CN molecule. Its excitation by a $\sim$100-fs pulse to the A state is followed by the direct dissociation monitored by probe pulses on-resonance and off-resonance with

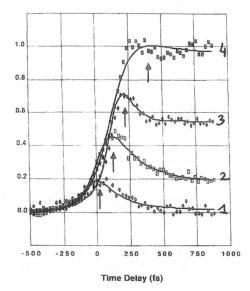

Time Delay (fs)

Fig. 5.14. The time dependence of the off-resonance (1–3) and on-resonance probe signal upon the excitation of ICN A-state (from [96])

absorption of the free CN fragment. The on-resonance probe (trace 4 in Fig. 5.14) shows a monotonic growth of the number of free CN fragments with the build-up time, t_{rise}, varying with excitation frequency in the 160-205-fs limits. The off-resonance absorption (traces 1–3) monitors CN still interacting with I: instead of oscillations, a single peak observed for each ω_{probe} value reflects the displacement of the wave packet from r_0 to $r \to \infty$ (Fig. 5.14).

Predissociation of Alkali Halides

The key experiment was realized for sodium iodide NaI [97]. The curve-crossing scheme is represented in Fig. 5.15: the potential curve of the diabatic, strongly bound ionic ground state, $X^1\Sigma_0^+$, with a high $Na^+ + I^-$ (42,830 cm^{-1}) dissociation limit, crosses the shallow potential curves of very weakly bound $A^1\Pi(0^+)$ and A' $^1\Pi(1)$ states correlated to $Na(^2S_{1/2}) + I^*(^2P_{1/2})$ ($E = 24,200$ cm^{-1}) and $Na(^2S_{1/2}) + I(^2P_{3/2})$ ($E = 17,200$ cm^{-1}) pairs. The $A-X$ interaction gives rise to a pair of adiabatic states: the excited $|1\rangle$ state covalent at $r \approx r_{eq}$ but tending to the $Na^+ + I^-$ limit and the ground $|2\rangle$ state ionic in the $r \approx r_{eq}$ range but correlated to $Na + I^*$. The A' $^1\Pi(1)$ and $X^1\Sigma_0^+$ curves cross without interaction in view of the $\Delta\Omega = 0$ Kronig rule.

The principle of the experiment is represented in Fig. 5.15. The jet-cooled NaI molecules are excited by the $\sim$60-fs light pulses with so large spectrum that both the A and A' states are populated. The fluorescence induced by the 150-200-fs probe pulses with variable delays and frequencies in the 16,300 to 17,400 cm^{-1} range containing the 16,950 cm^{-1} frequency on-resonance with Na $^2P_{1/2,3/2} \leftarrow$ $^2S_{1/2}$ transitions is

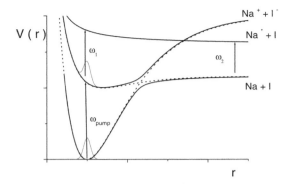

Fig. 5.15. The potential curves of NaI. The pump ω_{pump} and probe: off-resonance, ω_1, and on-resonance, ω_2, transitions are indicated by arrows

detected. The on-resonance pulse monitors the number of Na $^2S_{1/2}$ atoms formed by dissociation, whereas the off-resonance pulses probe the population of the excited (NaI)* state at different interatomic distances.

The results are shown in Fig. 5.16. The off-resonance signal oscillates with a ~1-ps period, as in the case of bound state, but its amplitude decreases at a few picosecond scale. The on-resonance signal shows an initial growth to about a half of its final value followed by a further stepwise increase with the step-to-step delay equal to the oscillation period. The initial growth is assigned to a direct dissociation of the $A^1\Pi(1)$ state excited above its dissociation onset, whereas the further stepwise process corresponds to the decay of the $^1\Pi(0^+)$ state.

These results may be rationalized in terms of the sequential coupling scheme (cf. Sect. 1.4.3 and Fig. 1.8). The pulse excitation prepares a coherent superposition of $|1\rangle$ and $|2\rangle$ eigenstates equivalent to the bright state NaI*. This state is monitored by

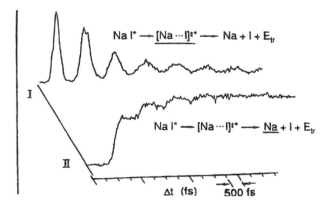

Fig. 5.16. On-resonance and off-resonance probe signals upon the electronic excitation of NaI (from [97]). Products probed by probe pulses are underlined

absorption corresponding to the Na* → Na** transition (the ionic Na$^+$I$^-$ state does not absorb in this spectral range). The observed beat period corresponds to the energy spacing $\Delta E_{12} \approx 6$ cm^{-1}. The free Na atoms are produced by dissociation of the unstable $^1\Pi(0^+)$ state, with a rate proportional to its population that oscillates with the beat frequency.

The same effect may be described in terms of the wave packet movement through the crossing of diabatic A and X surfaces (avoided crossing of adiabatic $|1\rangle$ and $|2\rangle$ surfaces) [97]. Dissociation of the (NaI)* molecule occurs when the wave packet follows in the crossing point the diabatic $\Pi(0^+$ potential curve of the A state. This trajectory corresponds to the jump $|1\rangle \rightarrow |2\rangle$ between adiabatic curves. Otherwise, the packet reflected on the potential barrier of the $|1\rangle$ state continues its oscillatory movement in the U_1 potential well. The probability P of the jump, estimated from the decay rate of the off-resonance signal, is of the order of 0.1 and varies slowly with the excitation frequency. The decay time of the excited state, estimated in this way, is of 5.4 ps at $\omega_{\text{exc}} \approx 32,200$ cm^{-1} and is compatible with 2.3-GHz widths of lines in the fluorescence spectrum of NaI in the $\sim 31,000$ cm^{-1} range [98].

5.4.4 Fast Relaxation of Polyatomics

The Geometry of Surface Intersections

The topology of multidimensional potential surfaces of polyatomic molecules is much more complex than in the case of diatomics [99, 100] and will not be treated here. There is, however, a close analogy with the crossing of one-dimensional potential energy curves, which may be illustrated in the case of nonlinear triatomic XY$_2$ molecule belonging to the C_{2v} symmetry group. In the harmonic approximation and at a fixed value of the $\theta_{\text{Y-X-Y}}$ valence angle, the potential surfaces U_A and U_B of $|A\rangle$ and $|B\rangle$ electronic states are the paraboloids

$$U_A(Q_a, Q_b) = f_a^A(Q_a - \widehat{Q}_a^A)^2 + f_b^A(Q_b - \widehat{Q}_b^A)^2 \qquad (5.35)$$

$$U_B(Q_a, Q_b) = f_a^B(Q_a - \widehat{Q}_a^B)^2 + f_b^B(Q_b - \widehat{Q}_b^B)^2 + \Delta U, \qquad (5.36)$$

where Q_a and Q_b are normal coordinates of the symmetric (A_1) and antisymmetric (B_2) X-Y stretching modes, f are the force constants, $\widehat{Q}$, the equilibrium values of Q coordinates and $\Delta U = U_B(\widehat{Q}) - U_A(\widehat{Q})$ is the electronic excitation energy of the $|B\rangle$ state. Note that $\widehat{Q}_b^B \equiv \widehat{Q}_b^A$ when the symmetry is conserved in the excited state. We will also assume that $f_b^B \approx f_b^A$, as it is often the case. The $U_B(Q_b)$ and $U_A(Q_b)$ parabolas have identical shapes and never cross. The geometry of the surface intersection depends on the relation between $\widetilde{Q}_a$ and f_a parameters in the $|A\rangle$ and $|B\rangle$ states: the surfaces cross when

$$f_a^B(Q - \widetilde{Q}_a^B)^2 + \Delta U = f_a^A(Q - \widetilde{Q}_a^A)^2.$$

One can differentiate two types of intersection events analogue of the glancing and peaked crossing of one-dimensional potential curves.

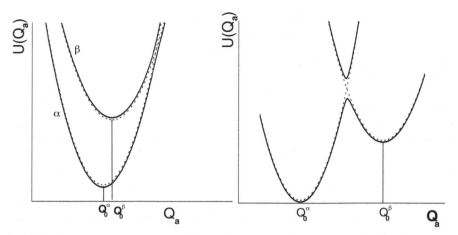

Fig. 5.17. Section through the $U(Q_a, Q_b)$ paraboloidal surfaces in the $Q_b = 0$ plane. Sloped (left) and conical (right) surface intersection

- The *sloped* surface intersection—a close analogue of the glancing curve crossing occurs when the equilibrium configurations in both states are not very different,

$$\widehat{Q}_a^A \approx \widehat{Q}_a^B.$$

The U_B paraboloid is then contained within the U_A surface. The crossing occurs when the surface of the excited state is shallower than that of the ground state. i.e., when $f_B^a < f_A^a$. The slopes of diabatic surfaces in the vicinity of the crossing point are not very different and adiabatic surfaces are nearly parallel (Fig. 5.17 (left)),

$$dU_B/dQ_a \approx dU_A/dQ_a.$$

The probability of transition between adiabatic surfaces is small, according to the Landau–Zener formula (Eq. 5.27).
- In the case of a large difference between equilibrium configurations of A and B states (Fig. 5.17 (right)) and in that of the crossing of bound and dissociative states, the intersection of surfaces is *peaked (conical)*. The slopes dU_B/dQ_a and dU_A/dQ_a are significantly different: the adiabatic surfaces form a double cone (diavolo) with the elliptical basis and a common apex. According to the Landau–Zener model, the probability of the $B \rightarrow A$ transition at the diabolical [98] conical intersection is much higher than in the sloped case.

The dynamics in the surface intersection region depends not only on their shapes but also on the strength of the A-B coupling. When the A and B states differ by their orbital symmetry, the coupling is forbidden in the equilibrium configuration and allowed only when the molecule is deformed, i.e., upon the excitation of a nontotally symmetric *promoting* mode. The crossing of the $U_A(Q_a)$ and $U_B(Q_a)$ parabolas in the $Q_b = 0$ plane is not a sufficient condition for an efficient relaxation. In order

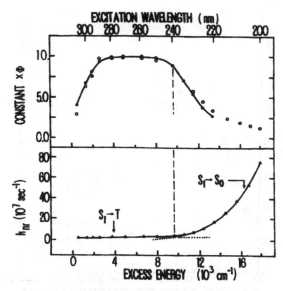

Fig. 5.18. Energy-excess dependence of the triplet yield (top) and of the S_1 nonradiative decay rate (bottom) of naphthalene

to attain the $Q_b \neq 0$ point, which constitutes a gate between two electronic states, a supplementary energy must be injected into the promoting mode. If this mode is not directly excited, the relaxation rate will be limited by population transfer from the initially excited level to the gate, which may be visualized by migration of the wave packet on the B-state surface. One can thus expect different relaxation rates of different levels with the same (or nearly the same) vibrational energy content.

We will discuss in a more detailed way the relaxation of high vibrational levels of the lowest electronic states of aromatics for which detailed information is available. We will tentatively apply to this case the model of the sloped intersection. We will rapidly resume the data concerning the ultrafast relaxation between higher electronic states of polyatomics. The conical intersection is generally considered as responsible for these processes.

The Third Relaxation Channel

Relaxation of High $S_1^{\#}$ Levels

A sudden increase of the nonradiative relaxation rate above a well-defined vibrational energy threshold was first observed in the S_1 state of benzene [101]. Its fluorescence yields and lifetimes vary slowly with increasing E_{vib} in a wide energy range: $\tau_f = 87$, 82 and 72 ns and $Q_f = 0.27$, 0.24 and 0.19 for $v = 0$, 1 and 2, the levels of the main vibrational progression: $6_0^1 1_0^v$ ($E_{\text{vib}} = 523 + 922v$ cm^{-1}). A sudden drop occurs for $v = 3$ and for other levels with $E_{\text{vib}} \geq 3,000$ cm^{-1}: $\tau_f < 3$ ns and $Q_f < 5 \times 10^{-4}$ [102].

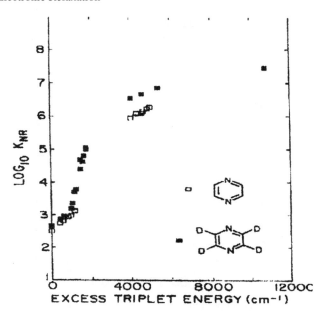

Fig. 5.19. The energy dependence of the relaxation rate of pyrazine and pyrazine-d_4 vapors (from [110])

A similar behaviour was observed for of a whole class of aromatic compounds, such as fluorobenzene [103], toluene and aniline [104] with onsets at about 2,500, 2,200 and 3,500 cm^{-1}, respectively. The third-channel onset increases with the size of the molecule (it amounts to ~9,500 cm^{-1} for naphthalene [105, 106]) and may be correlated with the average vibrational energy per mode rather than with the overall vibrational energy.

The low levels of the S_1 state decay by the fluorescence emission and the intersystem crossing $S_1 \rightarrow T_i$ with the yields $Q_f + Q_{ISC} \approx 1$. The role of the $S_1 \rightarrow S_0$ internal conversion is limited in view of the large $S_1 - S_0$ energy gap. The sudden drop of the fluorescence yield is not due to an increase of the k_{ISC}, the triplet yield being reduced in the same spectral range [107] as shown for naphthalene [105] (Fig. 5.18). Hence, the term of "the third relaxation channel" for this process, which cannot be explained either by photochemical reactions and was assigned to the sudden enhancement of the internal conversion. This assignment is now generally accepted in spite of the absence of direct proof that high $S_0^{\#}$ levels are populated.

Relaxation of High $T_1^{\#}$ Levels

A strong E_{vib} dependence of decay rates is also observed for high $T_1^{\#}$ levels of benzene, toluene and pyrimidine [107–109], the most complete data being obtained for pyrazine upon the direct excitation of single T_1 vibronic levels in a wide E_{vib} range [110, 111].

The steep increase of the $T_1 \rightarrow S_0$ intersystem crossing rate from 4.6×10^2 s^{-1} for $E_{vib} = 0$ to 1.1×10^5 s^{-1} for 1,750 cm^{-1} is followed by a quasi-saturation. The sigmoidal plot of k_{ISC} vs. E_{vib} represented in Fig. 5.19 is different from the rapid acceleration of the $S_1^\#$ relaxation rate above the third-channel onset.

The Sloped-Intersection Model

The electronic spectra of aromatic and aza-aromatic molecules show that the differences between their equilibrium configurations, $\widehat{Q}$, in the S_1 and S_0 states are not very pronounced, whereas the frequencies of the totally symmetric a modes are significantly changed. For instance, that of the ring breathing mode of benzene is reduced from $\omega_1 = 992$ cm^{-1} in the S_0 state to 920 cm^{-1} in the S_1 state.

The $B \rightarrow A$ relaxation may be described in terms of paraboloidal surfaces involving a totally symmetric a mode and promoting b mode. In the case of the ground $^1A_{1g}$ and excited $^1B_{2u}$ states of benzene, the $\omega_1(a_{1g})$ and the ring-deforming $\omega_9(b_{2u})$ may be identified as a and b modes.

Several predictions of this model may be checked by experiment:

- The model predicts that high ground-state levels of the energy-accepting a mode are selectively populated by the $S_1^\# \rightarrow S_0^\#$ internal conversion. Their population may be probed by measurements of the hot absorption $S_0^\# \rightarrow S_i^\#$ after the pulse excitation but, to our best knowledge, this experiment has not yet been attempted.
- In the sloped intersection case, once a threshold of a rapid relaxation is attained, a further increase of the decay rate with E_{vib} is slower in view of nearly parallel surfaces of the A and B states. The saturation of k_{ISC} with the E_{vib} increase in the case of the pyrazine triplet state may be explained in this way but no data are available for the S_1 states of benzene and its derivatives.
- Because the $S_1 - S_0$ coupling is induced by promoting modes, one can expect that only a fraction of vibrational levels with the energy exceeding the intersection onset are strongly coupled to the $S_0^\#$ quasi-continuum and rapidly depopulated. The decay rates of other states are determined by the intrastate coupling to these gate states. The validity of this statement was confirmed in a study of the decay rates of the individual ro-vibronic levels of benzene in the third-channel energy range.

The High-Resolution Spectroscopy of Benzene

The scope of this experiment was to show the perturbations of the rotational structure of bright vibrational levels of the S_1 state in the third-channel region due to their coupling to short-lived gate states.

The rotational structure of the $^1B_{2u} \leftarrow {}^1A_{1g}$ absorption has been resolved by the sub-Doppler spectroscopy of room-temperature vapors using a two-photon excitation by two anticollinear circularly polarized laser beams (cf. Sect. 2.2.2). The spectral resolution of 5 MHz for a continuous (cw) excitation and of 70 MHz for ~20-ns light pulses was attained, allowing the measurements of level widths corresponding to decay rates $k_s \geq 5 \times 10^6$ s^{-1} [112]. Only the ro-vibronic states of the E_{1g} overall

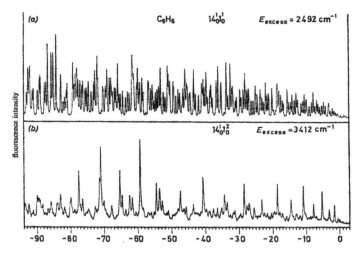

Fig. 5.20. The rotational structure of the vibronic transitions in the fluorescence spectrum of benzene vapors below and above the third-channel onset (from [112])

symmetry are bright upon the two-photon excitation. They are mixed with dark $|\ell\rangle$ states of a different symmetry by the Coriolis effect: the coupling to E_{2g} levels is induced by the in-plane (a_{2g}) rotation component and to the A_{1g}, A_{2g}, E_{2g} ones by its out-of-plane (e_{2g}) component. The coupling strength is proportional, respectively, to K and $\sqrt{(J^2 - K^2)}$ (cf. Sect. 3.5). The perturbations are observed even in transitions

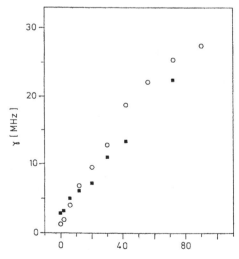

Fig. 5.21. The J dependence of widths of rotational J, $K = 0$ levels of the benzene $14^1 1^2$ state (from [113])

to low vibronic states with energies below the third-channel onset (Fig. 5.20 (a)): about 10% of the rotational levels of the 14^1 state ($E_{vib} = 1571$ cm^{-1}) are shifted and/or split but this perturbation has no effect on the level widths (lifetimes). The widths of perturbed and unperturbed rotational lines of the 14^1 and still higher $14^1 1^1$ ($E_{vib} = 2,489$ cm^{-1}) and $14^1 1^1 16^1$ ($E_{vib} = 2,726$ cm^{-1}) states are the same, of the order of 10 MHz, in good agreement with directly measured lifetimes in the 120–134-ns range [113]. The perturbation mixes the states with the same (or nearly the same) homogeneous widths.

In the third-channel region, the effects are much more spectacular. The rotational structure of the transition to the $14^1 1^2$ ($E_{vib} = 3,412$ cm^{-1}) state is limited to a few lines corresponding to J', $K' = 0$ levels (Fig. 5.20 (b)); all the $K \geq 1$ levels are not fluorescent [114]. The widths of the $K' = 0$ lines (corrected for instrumental and pressure broadening) increase from 1.3 MHz for $J = 0$ to 46.1 MHz for $J = 14$ that corresponds to k^{nr} varying as $J(J + 1)$ from 8.2×10^6 to 2.9×10^8 s^{-1}, in good agreement with directly measured decay times (Fig. 5.21) [115]. The perturbation of a lower $14^1 1^1 16^2$ $E_{vib} = 2,963$ cm^{-1}) state is less pronounced: several lines of transitions to $K' \geq 1$ levels persist but are broadened and their widths increase with K'.

Similar effects are observed for vibronic levels with a different symmetry recorded upon one-photon excitation: e.g., the $6^1_0 1^3_0$ band ($E_{vib} = 3,287$ cm^{-1}) is so strongly perturbed that the rotational analysis could not be done. The homogeneous widths of individual spectral features vary between 150 and several hundred MHz, in agreement with directly measured decay times in the $\tau < 1$-ns to $\tau \approx 7$-ns limits [116]. Unfortunately, no data are available for the excited singlet levels with $E_{vib} \gg 3,500$ cm^{-1}, well above the third-channel onset.

These results may be rationalized in terms of the sequential Coupling, $|s\rangle \leftrightarrow |\ell\rangle \leftrightarrow \{m\}(S_0^{\#})$, states. Above the third-channel threshold, several $|s\rangle$ levels are still fluorescent but relax with a J, K-dependent rate to dark ℓ levels, non-radiant because they are very strongly coupled to the continuum of $S_0^{\#}$ states. The relaxation rate is limited by that of the slow $s \rightarrow \ell$ step. In the wave-packet language, this picture corresponds to a migration of the wave packet created on the $U_B(Q)$ surface until it arrives to the gate in the $U_B(Q) - U_A(Q)$ intersection ridge.

Ultrafast Relaxation of Polyatomics

Actual studies of ultrafast processes in complex systems have to be considered as the initial steps on a steep path. The theoretical treatment of the events occurring on the multidimensional surface necessitates either very heavy calculations or a drastic reduction of dimensions of the system to a few normal coordinates. The interpretation of experimental data is uncertain in view of large homogeneous widths of overlapping absorption and emission bands and nonexponential decay curves. A nonsatisfactory agreement between calculated and observed spectra or rates is current and not surprising.

The effect of the surface intersection on the absorption spectra was calculated for the model potentials [100, 116–117], but the fine structure of simulated spectra is blurred out in view of the short deactivation times and large widths of its components. For instance, the calculated absorption of the butadiene ^1B_u state contains a large number of closely spaced resonances, whereas the observed spectrum is composed of a few diffuse bands with $\delta\omega \approx 500$ cm^{-1} homogeneous widths corresponding to $\tau \approx 15$ fs [118].

The experimental and theoretical effort was concentrated on a few model systems: the $S_1 \rightarrow S_0$ relaxation of azulene [119, 120], the $S_2 \rightarrow S_1$ relaxation in pyrazine [121–123] and the $S_2 \rightarrow S_1 \rightarrow S_{01}$ processes in polyenes [124–132]. We will briefly discuss only the last of them.

Polyenes

The case of polyenes $C_{2n}H_{2n+2}$ ($n \geq 2$) is interesting because of the dependence of the relaxation paths on the length of the -C=C- chain. Their common features are the $\pi \rightarrow \pi^*$ absorption spectra composed of more-or-less overlapping strong $1A \rightarrow 1B$ and weak (forbidden in the C_{2h} symmetry group) $1A \rightarrow 2A$ transitions [130]. In spite of the large oscillator strength of the $1A \leftrightarrow 1B$ transition, no emission from the initially excited $1B$ state was observed. Moreover, the $2A$ excited state of short ($n = 2$ and 3) polyenes are also nonfluorescent, whereas octatraene ($n = 4$) shows an intense long-lived fluorescence in the gas and condensed phases, with a 250-ns lifetime in n-octane matrices at 4 K [132]. Such a long lifetime is consistent with the weakness of the $1A \rightarrow 2A$ absorption but shows also that the $2A \rightarrow 1A$ internal conversion is inefficient. The fluorescence is also observed for longer polyenes [124].

The absence of the $1B \rightarrow 1A$ fluorescence is obviously due to the ultrafast $1B \rightarrow 2A$ ($S_2 \rightarrow S_1$) relaxation. Its rate, roughly estimated from resonance Raman spectra and absorption bandwidths, amounts to 2×10^{13} and 5×10^{13} s^{-1} for the isomers of hexatriene ($n = 3$). The 2×10^{13} and 10^{14} s^{-1} decay components observed for the *trans*-butadiene ($n = 2$) in the MPI experiment are assigned to the $^1B_u \rightarrow {}^1A_g$) relaxation [124]. These rates are consistent with the conical intersection of $1B$ and $2A$ surfaces (Fig. 5.22) deduced from the *ab initio* calculations [129, 130].

The decay rates of the $2A$ state are strikingly different for short- and long-chain polyenes. This difference is probably due to the nonplanar equilibrium structure of the $2A$ states of butadiene and hexatriene, whereas those of $\pi^{n-1}\pi^*$ states of octatetraene and of higher members of polyene series are planar. It is well known [133] that equilibrium configuration of the $\pi\pi^*$ $^1B_{1u}$ state of ethylene is twisted with nearly perpendicular planes of CH_2 groups. This structure, due to the strong antibonding character of π^* electrons, implies the conical intersection of the ground and excited surfaces and an ultrafast electronic relaxation. The *ab initio* calculations suggest that the $2A$ and $1A$ surfaces intersect also in butadiene [127] and in all-*trans*-hexatriene (Fig. 5.22) [128].

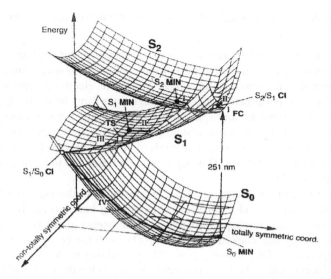

Fig. 5.22. Calculated surfaces of S_2, S_1 and S_0 states of hexatriene (from [128])

5.5 Conclusions

The wide difference between coupling patterns of isoenergetic vibrational levels of different electronic states is the origin of a large diversity of relaxation processes. They are distributed between two poles:

- the periodic time evolution resulting from the coherent excitation of a set of discrete levels in the absence of intra- or intermolecular dissipative continua (the small-molecule limit). The decay rate is determined by radiative rates of the relevant states and the quantum yield of emission is equal to one,
- the irreversible exponential decay with the $k_s = k_s^{rad} + k_s^{nr}$ rate and the $k_s^{rad} / k_s \le 1$ fluorescence yield resulting from a direct coupling of the bright $|s\rangle$ state to the dissipative continuum $\{m\}$.

A finer analysis of experimental data shows that nearly all real systems belong to an intermediate zone in which the coupling between discrete levels as well as their coupling to the continuum must be taken into account. These systems may be schematically described in terms of the sequential coupling: $s \leftrightarrow \ell \leftrightarrow \{m\}$. This picture is still oversimplified, the tier model, developed for description of IVR (cf. Sect. 3.6), seems to be more adequate but is still very seldom applied to the electronic relaxation. It may be useful in the treatment of the time evolution of small and intermediate-sized molecules perturbed by interactions with their environment (transition from the small- to the large-molecule limit).

The electronic relaxation rate varies in extremely large limits from k_{rel} of the order of 1 s^{-1} for $T_1 \rightarrow S_0$ transitions in aromatics to 10^{13}–10^{14} s^{-1} rates of internal conversion between excited singlet states. We consider that it is convenient to separate two types of processes:

- The systems in which the potential-energy surfaces of A and B adiabatic states are well separated at least in the fraction of the configuration space containing their equilibrium structures so that $\Delta U_{AB}(Q) \geq \omega_A$. The electronic relaxation is then induced by weak perturbations due to the deviations from the Born–Oppenheimer approximation (vibronic and spin-orbit coupling). For a given family of compounds for which the electronic coupling strength is roughly the same, the relaxation rate is limited by the overlap integrals of vibrational wave functions $\langle \chi_A | \chi_B \rangle$ of the initial and final states. This dependence is described in terms of energy- or momentum-gap laws.
- the Born–Oppenheimer approximation breaks down in the vicinity of the intersection of diabatic a and b surfaces. The coupling between adiabatic A and B states is no more a weak perturbation and the energy-gap law is meaningless when the energy gap goes to zero: $\Delta U_{AB}(Q) \to 0$ for $Q \to Q_c$, where Q_c is the configuration of the crossing point. The probability of the $A \leftrightarrow B$ transitions depends (according to the Landau–Zener model) on the difference of $(\partial U / \partial Q)_c$ slopes, so that the relaxation is more rapid in the case of conical than in that of the sloped intersection.

The sensitivity of intramolecular processes to environment effects depends on their time scale: the slow relaxation will be strongly influenced by a rapid reorientation of their solvation envelope, whereas the solvent movements are too slow to modify the processes occurring at the femtosecond scale.

6

The Electron and Proton Transfer

6.1 General Remarks

We will treat in this chapter the relaxation pathways intermediate between the purely photophysical relaxation (e.g., $S_i \rightarrow S_j$ or $S_i \rightarrow T_j$ transitions) and monomolecular photochemical reactions: the closely related processes of *electron transfer* and *proton (or hydrogen-atom) transfer*.

The electron transfer (redox reactions) and proton transfer (acid–base reactions) play a fundamental role in chemical and biochemical processes. The recent studies carried out for a large variety of molecular systems and environments are reviewed in a number of papers (cf. [1–5] and the special issue of Adv. Chem. Phys **106–107**, (1999) for the electron transfer and [6–8] for the proton transfer).

The photo-induced electron and proton transfer processes have been studied mainly in the condensed phases and they are extremely sensitive to environment effects because of a large difference between dipole moments of neutral and ionic states, which implies a strong solvent dependence of their stabilization energies. It is practically impossible to separate the parameters of the molecular system and of the bath; this coupling is taken into account in the Marcus theory [9]—the basis of the theoretical treatment of electron-transfer processes in condensed phases. We are, however, interested in their intramolecular aspects and focus our attention on experiments involving isolated molecules and molecular complexes or, at least, the systems only slightly perturbed by their environment.

Because the key problems appear already in isolated molecules, we use the formalism of nonradiative processes and refer only occasionally to that of Marcus. We apply the concepts developed in previous chapters, such as diabatic and adiabatic surfaces, surface intersections and their role in the electronic relaxation processes. We will often switch from the basis of adiabatic states to that of diabatic states and *vice versa*.

We use the term electron transfer for a process in which the electronic charge distribution is so strongly modified that this process may be described in terms of transitions between *diabatic* neutral and ionic states of A + D:

$$A..D \leftrightarrow A^-..D^+$$
$$A^*..D \leftrightarrow A^-..D^+$$
$$A..D^* \leftrightarrow A^-..D^+,$$

where the *electron-acceptor* and *electron-donor* units A and D are either two molecules (free or forming a complex AD) or two groups within an A-B-D molecule, where -B- is an aliphatic chain or a set of conjugated (π-electron) bonds. A..D is the *ground*—$|G\rangle$ state of the system, A^*..D and A..D^* are its *locally excited*—$|LE\rangle$ states and A^-..D^+ is the *charge transfer*—$|CT\rangle$ state. This process may also be described as a displacement of the electronic charge taking place in one of the *adiabatic* states of A + D: the ground $|A\rangle \approx |G\rangle$ or excited $|B\rangle = \alpha|LE\rangle + \beta|CT\rangle$ state. In the first approximation, all higher electronic states such as diabatic ionic A^+..D^- and adiabatic $|B'\rangle = \beta|LE\rangle - \alpha|CT\rangle$ states are ignored.

A direct consequence of a change of the electronic charge distribution may be a displacement of nuclei to a new equilibrium configuration. In the special case of an A..HD and A-DH..C systems involving an intra- or intermolecular hydrogen bond, this process is described as the proton (or hydrogen atom) transfer. A large shift of the electron charge

$$A..HD \rightarrow A^{-\delta e}..HD^{+\delta e}$$

(where e is the electron charge and $0 \leq \delta \leq 1$) strongly modifies the electric field acting on the proton and induces its displacement:

$$A^{-\delta e}..HD^{+\delta e} \rightarrow AH^{+(1-\delta e)}..D^{-(1-\delta e)}.$$

When $\delta e \approx 1$, the transfer of the proton charge is compensated by that of electronic charge and the process corresponds to the transfer of the neutral atom (the *H-atom transfer*),

$$A..HD \rightarrow AH..D,$$

whereas, in the $\delta e \ll 1$ limit, the process may be described as a *proton transfer* with formation of a hydrogen-bonded *ion pair (IP) state*,

$$A..HD \rightarrow AH^+..D^-,$$

A..HD and AH^+..D^- states are diabatic states of the system. The same process may be described as the proton movement on a single adiabatic surface with a pronounced dependence of the electronic wave function on the nuclear coordinates. Pure proton and hydrogen-atom transfer correspond to the limiting cases of such surfaces.

We focus our attention on a limited class of processes involving the electronic relaxation of locally excited $|LE\rangle$ states via transition to the $|CT\rangle$ state (*charge separation*) followed by a transition to the ground state by a reverse electron transfer (*charge recombination*),

$$A^*..D \rightarrow A^-..D^+ \rightarrow A..D$$

or by a two-step proton transfer:

$$A^*..HD \rightarrow AH^+..D^- \rightarrow A..HD.$$

These relaxation channels are open when the energies of the ionic $|CT\rangle$ or $|IP\rangle$ states are lower than those of corresponding $|LE\rangle$ states at least in a part of the configuration space.

The essential parameters for electron-transfer processes are the ionization potential of the donor (I_D) and the electron affinity of the acceptor ($\mathcal{E}_A$). At an infinite A..D distance ($R_{AD} \rightarrow \infty$), the potential energy of the $|CT\rangle$ state is that of a free ion pair, $A^- + D^+$, and that of the $|LE\rangle$ state, $A^*..D$, is the excitation energy of A^*,

$$U_{CT}(R_{AD} = \infty) = I_D - \mathcal{E}_A, \quad U_{LE}(R_{AD} = \infty) = E_{A^*}. \tag{6.1}$$

For typical aromatic donors and acceptors, I_D varies in the 7–10.5 eV range whereas $\mathcal{E}_A \approx 0.5$–1.5 eV and $E_{A^*} = 3$–5 eV, so that $U_{CT}(\infty) \gg U_{LE}(\infty)$. The energy of the CT state is, however, more strongly reduced for $R_{AD} \rightarrow 0$ by the Coulomb interaction than those of LE and G states, weakly bound by the van der Waals forces. In the first approximation of a one-dimensional surface, $U_i(R_{AD})$,

$$U_{CT}(R_{AD}) = I_D - \mathcal{E}_A - C(R_{AD}), \tag{6.2}$$

where $C(R_{AD}) = -e^2/R_{AD} \approx -14.4/R_{AD}$ in eV/Å units, while the potentials of LE and G states are decreased by $\Delta U \approx 0.1$–0.25 eV only so that, at $R_{AD} \approx 3.5$ Å, the energies of LE and CT states are not very different (Fig. 6.1 (a)). The electron transfer from acceptor to donor, $A..D \rightarrow A^+..D^-$, is also possible but, because the energy of the $A^+..D^-$ state is significantly higher than that of the A^-D^+ state, its role in relaxation processes is limited.

Similar relations characterize the proton transfer with formation of an ion pair. At an infinite R_{AD} distance, the energy of the IP state $AH^+...D^-$ (bonding energy of H^+D^- minus the proton affinity of A) is usually larger than that of the neutral A..HD pair but the energy of the ion pair containing a coulombic term rapidly decreases when $R_{AD} \rightarrow 0$. In both types of systems, the potential curves of diabatic neutral LE and ionic (CT or IP) states intersect at the crossing distance, R_c, so that the potential curve of the adiabatic $|B\rangle$ state strongly deviates from the usual Morse form and may contain the energy barriers. Since the $U_{CT} - U_{LE}$ energy gap varies rapidly with R_{AD} the α and β mixing coefficients are R-dependent so that the properties of excited A..D and A..HD system vary strongly with R_{AD}. The interactions CT–G and IP–G involving the ground electronic state must also be taken into account.

The one-dimensional model $U(R_{AD})$ is obviously oversimplified. The energies of all states depend not only on the A..D distance but also on other coordinates, such as the mutual orientation of A and D groups.

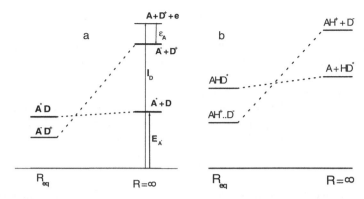

Fig. 6.1. Scheme of levels involved in (a) the electron transfer and (b) the proton transfer process

6.2 Electron Transfer

We will treat the $LE \rightarrow CT$ and $CT \rightarrow G$ electron transfer in two kinds of systems:

- the A + D pairs free or forming the A..D complexes in supersonic jets and crystals or solutions. The equilibrium geometries of their G, LE and CT states are determined uniquely by the dependence of the A–D interaction energy on the R_{DA} distance and the A..D mutual orientation.
- the bichromophores A–B–D in which A and D are linked by a flexible, semirigid or rigid aliphatic bridge B. Their equilibrium configurations depend not only on the A–D interaction but are limited by the structure of the bridge. Because the energies of excited $\sigma\sigma^*$ states of the bridge are much higher than those of the A*D and A$^-$D$^+$ states, the A–B and D–B interactions may be, in the first approximation, neglected. The strength of the A–D interaction, as compared with the free A + D pair is, however, enhanced by hyperconjugation effects and polarizability of the bridge [10].

We will only briefly mention the A–B–D systems in which the bridge B is formed by a system of π-electron (conjugated or aromatic) bonds. In this case, the diabatic A*–B–D or ionic A$^-$–B–D$^+$ are so strongly mixed with the excited A–B*-D states of the bridge that excitation is delocalized over the whole A–B–D system. We ignore also the charge redistribution processes in molecular ions A-B-D$^-$, which do not involve ionic and neutral states but only the electron migration, A–B–D$^-$ $\rightarrow$ A$^-$–B–D.

The structures of several model A..D and A-B-D systems are represented in Fig. 6.2.

6.2.1 Potential Energy Surfaces

The approximation of potential-energy surfaces of adiabatic, $U_A(Q)$, $U_B(Q)$, and diabatic, $U_G(Q)$, $U_{LE}(Q)$ and $U_{CT}(Q)$ states, by one-dimensional potential curves, $U(R_{AD})$, is good as long as R_{AD} is large as compared with the van der Waals radii,

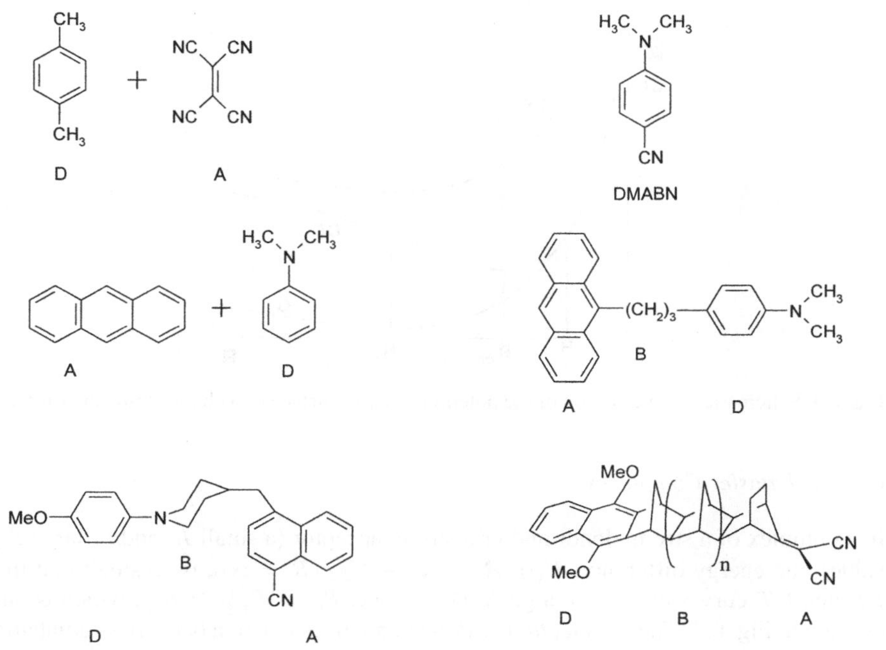

Fig. 6.2. Structural formulae of a few model electron donor–acceptor systems

d_A and d_D, of A and D molecules (groups). It breaks down in the case of a tight contact complex, the energy of which strongly depends on angular coordinates. We are essentially interested in a fraction of the coordination space containing equilibrium configurations of G, LE, CT and $|A\rangle, |B\rangle$ states. Information about this part of space may be deduced from absorption and emission spectra of the bound states of AD complexes.

Because

$$U_{CT}(R_{CT}^{eq}) < U_{LE}(R_{LE}^{eq}),$$

whereas

$$U_{CT}(R = \infty) \gg U_{LE}(R = \infty),$$

the $U_{CT}(R_{AD})$ and $U_{LE}(R_{AD})$ diabatic surfaces cross (Fig. 6.1 (a)). The shape of the $U_B(R)$ adiabatic surface depends on the position of the crossing point R_c with respect to the equilibrium distances in the CT and LE states, R_{CT}^{eq} and R_{LE}^{eq}. We will discuss this question in the simple case of free A..D pairs.

Free A..D Systems

We define the *charge-transfer complex* as a molecular system with $R_c > R_{LE}^{eq}$ and the *exciplex* as that with $R_c < R_{LE}^{eq}$ (Figs. 6.3 and 6.4). This definition is compatible with the current use of these terms: the ground-state of charge-transfer complexes is bound in inert solvents whereas the ground state of exciplexes is unstable in room-temperature gases and solutions.

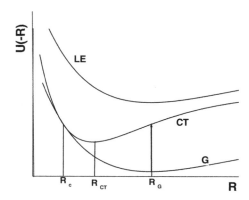

Fig. 6.3. Schematic representation of the potential-energy surface of a charge-transfer complex

Charge-Transfer Complexes

In a complex of a strong donor and/or a strong acceptor (a small I_D and/or large $\mathcal{E}_A$ value), the energy difference, $U_{CT}(R = \infty) - U_{LE}(R = \infty)$, is so small that the LE and CT curves cross at a large A..D distance, $R_c \gg R_{LE}^{eq} \approx R_G^{eq}$, which is out of scale in Fig. 6.3. The *intersection is sloped* and gives rise to a barrierless adiabatic U_B surface. Its energy minimum corresponds to the tightly bound *"contact" complex* with the equilibrium distance $R_B^{eq} \approx R_{CT}^{eq} \approx d_A + d_D$. The complex is strongly bound with R_B^{eq} significantly smaller than R_A^{eq}.

At $R \approx R_B^{eq}$, the $|B\rangle$ state has a predominant CT character with a small admixture of the LE character due to a large $U_{LE} - U_{CT}$ energy gap.

On the other hand, the interaction between the CT and G states closely spaced in the $R_{AD} \approx R_{CT}^{eq} \leq R_G^{eq}$ range is relatively strong. The adiabatic ground state, $|A\rangle = \alpha|LE\rangle + \beta|CT\rangle$, is stabilised by an admixture β^2 of the strongly bound CT state. This admixture is evidenced by the nonzero dipole moment, μ_A, of the complex of two apolar molecules (as e.g., benzene-iodine) borrowed from the CT state with $\mu \approx eR$,

$$\mu_A \approx \beta^2 \mu_{CT} \approx \beta^2 e R_G^{eq}.$$

The A–D bonding energy is thus larger than that of the purely van der Waals interaction of its components with the solvent C so that the concentration of AD complex in solutions is large enough to give rise to the $G \to CT$ (A..D $\to$ A$^-$..D$^+$) absorption [12]. Because, at $R_{AD} \approx R_A^{eq}$, CT is the lowest excited state ($U_{CT} < U_{LE}$), the optical excitation of the ground-state AD complex prepares directly the CT state.

The $U_{CT} - U_G$ energy gap is still reduced at $R_{AD} \leq R_{CT}^{eq}$ so that another sloped intersection of diabatic surfaces may occur in strongly bound complexes (Fig. 6.3).

Exciplexes

In the complexes of weaker donors or acceptors, the attractive part of the CT surface intersects the repulsive part of the LE surface at $R_c < R_{LE}^{eq}$ and *the intersection is*

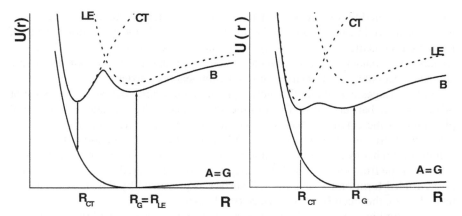

Fig. 6.4. Schematic representation of the potential energy surface of an exciplex with and without an energy barrier

peaked (Fig. 6.4). This implies an energy barrier on the adiabatic surface, $|B\rangle$ the height of which depends on the energy of the crossing point of diabatic surfaces, $U(R_c)$, and the strength of the LE–CT coupling in this point, $V_{LE,CT}(R_c)$,

$$U_B(R_c) = U(R_c) - V_{LE,CT}(R_c).$$

In the case of a strong coupling, the energy of the adiabatic state is lowered in the R_c region so strongly that the energy barrier is suppressed and replaced by an inflection of the $U_B(R)$ curve.

The evolution of the A*D system prepared in the $|B\rangle$ state with a predominant LE character by an A* + D collision or by vertical excitation of an AD complex at $R_{AD} \approx R_G^{eq}$ depends on the surface shape. As in the case of charge-transfer complexes, the equilibrium configuration of the contact complex has a predominant CT character but the admixture of LE is nonnegligible. On the other hand, the CT–G interaction is weaker and, in the absence of a significant CT admixture, the $|A\rangle$ ground state is bound only by van der Waals forces. The solvated AD_s ground-state complex in a room-temperature solution is not more stable than the $A_s + D_s$ free pair, but cold AD complexes are formed in supersonic expansions.

A–B(σ)–D Systems with Aliphatic Bridges

The structure of the ground electronic state corresponds to the equilibrium configuration of the bridge [13, 14]; the A..D interaction in the ground state of typical A–B–D systems is practically absent, as evidenced by electronic absorption spectra of jet-cooled species: the $G \rightarrow LE$ (A $\rightarrow$ A*) absorption band of A-B-D molecules is nearly the same as that of A–B [15].

The oligomethylene bridge (-CH$_2$-)$_n$ ($n \geq 3$) is flexible: the energy minima corresponding to trans, cis and gauche configurations of the chain have similar depths and are separated by low-energy barriers. In the room-temperature gas, several isomeric forms are in equilibrium but the *all-trans* is probably the predominant form of jet-cooled molecules [15]. The energy minimum of the CT state A^--B-D^+ corresponds to the contact complex between A^- and D^+ groups, i.e., to the folded structure B-chain, necessitating a transition across the low-energy barriers.

In the bridges containing the more rigid aliphatic rings, such as cyclohexane or piperidine, the barriers between conformers are higher [13]. The stable structure of piperidine is the extended chair structure (cf. Fig. 6.2), whereas the energy of the folded boat structure with a reduced A..D distance allowing formation of the contact complex is higher by $2,000\,\mathrm{cm}^{-1}$ and separated by the energy barrier of $\sim 3,500\,\mathrm{cm}^{-1}$. The transition from the extended structure of A^*-B-D prepared by optical excitation to the folded form on the shallow LE surface is prevented by this barrier but may take place on the steep $U_{CT}(R)$ surface (see below).

In the limiting case of a rigid B spacer, such as the norbornane bridge (Fig. 6.2) [10], the barrier is so high that the configuration of the system determined by the spacer geometry is the same in G, LE and CT state.

Conjugated A–B(π)–D Systems

When the donor and acceptor groups are bound by chains of conjugated bonds or aromatic rings (Fig. 6.2) the excited adiabatic $|B\rangle$ state is a superposition of the A^--B-D^+ and AB^*D states. The molecule is strongly polar [16] but the dipole moment is smaller than expected in the pure ionic A^-BC^+ structure with complete charge separation.

Several systems of this type (such as DMABN, dimethylaminobenzonitrile, represented in Fig. 6.2) show in polar solvents the emission that suggests a complete charge separation and is explained by formation of an isomeric *intramolecular charge transfer ICT* state in which the D^+ (dimethylamino) donor group is decoupled from the A^--B (benzonitrile) acceptor part of the molecule. This decoupling may correspond to an out-of-plane configuration of the molecule in the *twisted intramolecular charge transfer, TICT* state [17]. Other structures of the ICT state have been proposed. For a critical review of the recent works, cf. [18–20] and especially the recent and extensive treatment of theoretical and experimental data in [21]. Because the ICT states have never been observed in the isolated model molecules but are induced by solvent effects, we will not treat this subject here.

6.2.2 Spectroscopy and Dynamics

We will start by a discussion of the mechanism of the $CT \rightarrow G$ relaxation, which is the same in the case of CT states prepared by a direct $G \rightarrow CT$ optical excitation of charge transfer complexes and by the $LE \rightarrow CT$ relaxation of exciplexes.

$CT \leftrightarrow G$ Transitions

The essential features of the $|A\rangle \leftrightarrow |B\rangle$ radiative and nonradiative transitions result from a large difference between the shapes of the potential surfaces, $U_B \approx U_{CT}$ and

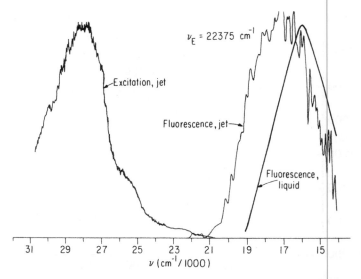

$\nu_E = 22375 \ cm^{-1}$

Excitation, jet

Fluorescence, jet

Fluorescence, liquid

$\nu \ (cm^{-1}/1000)$

Fig. 6.5. Fluorescence excitation and fluorescence spectrum of the p-xylene–TCNE complex in the supersonic jet and in solution (from [23])

$U_A \approx U_G$, and especially between the equilibrium distances, $R_{CT}^{eq} \ll R_G^{eq}$ (Fig. 6.3). The probabilities of radiative transitions between G and CT states are low because of a small overlap between the donor orbital of D and the acceptor orbital of A. The $G \rightarrow CT$ absorption bands of charge-transfer complexes are thus relatively weak and the radiative lifetimes of CT states relatively long. The vertical transitions from the equilibrium configurations of the initial state attain a dense set of high vibrational levels of the final state in absorption as well as in emission. The spectra differ from those of aromatic molecules and of their van der Waals complexes by a large emission-to-absorption Stokes shifts, $\Delta\omega_{St} \approx 7\ 000 \ cm^{-1}$, and the absence of the vibrational structure [22, 23]. An electronic transition gives rise to a single, broad and diffuse band ($\delta\omega \approx 3,000–5,000 \ cm^{-1}$) structureless even in the gas-phase [24, 25] and in supersonic expansions [23]. The absence of a significant narrowing of the emission spectrum in the jet-cooled complex with respect to solution (Fig. 6.5) indicates that its bandwidth is the intrinsic property of the system

The broadening of the high vibrational levels of the ground electronic state may be due to the vibrational predissociation: the Stokes shift is larger than the bonding energy of AD in its $|G\rangle$ state of the order of 2,000 cm^{-1} [12], so that fluorescence populates the levels above the dissociation onset.

This argument cannot be applied to the absorption spectra, the CT state being strongly bound. Their large widths may be due to a rapid reorganization of the excited complex from the initially prepared Franck–Condon configuration to the equilibrium configuration. This process is responsible for the frequency shift of the CT emission of the HMB-TCNE, hexamethylbenzene-tetracyanoethylene [26] (Fig. 6.6) and of the transient absorption of the pyrene-TCNE complex [27, 28] in nonpolar solvents

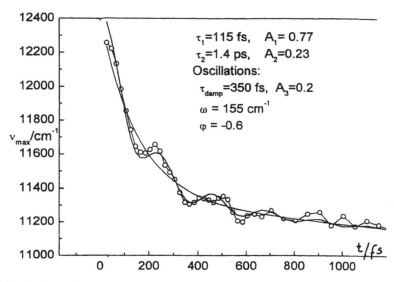

Fig. 6.6. The time dependence of ω_{max} of the CT fluorescence of the HMB-TCNE complex (from [26])

with the $k \approx 10^{13}$ s^{-1} rate, which implies a homogeneous broadening of vibrational levels of the order of 50 cm^{-1}.

The fluorescence spectra of *exciplexes* emitted from their CT states are broad and structureless like those of charge-transfer complexes, but the energies of their CT states are higher, so that, for $R_{AD} \approx R_{CT}^{eq}$, the $U_{LE} - U_{CT}$ energy gap is smaller and the $U_{CT} - U_G$ larger than in charge-transfer complexes. This difference is a probable reason of the striking difference between the efficiency of the nonradiative decay channels in charge-transfer complexes and exciplexes.

The radiative decay rates, k_{CT}^{rad}, of CT states are nearly the same (of the order 10^6 s^{-1}) for the benzene-TCNE charge-transfer complex [22] and for the anthracene–dimethylaniline exciplex [29]. Surprisingly, their nonradiative rates differ by many orders of magnitude. The yields and lifetimes of the exciplex fluorescence are large and surprisingly insensitive to temperature and environmental effects. In the case of the anthracene–dimethylaniline complex, k_{CT}^{nr} is practically the same in the supersonic expansion ($\sim 10^6$ s^{-1}) [30] as in the room-temperature cyclohexane solution (1.3×10^6 s^{-1}) [29]. In contrast, the fluorescence of charge-transfer complexes is extremely sensitive to temperature and solvent effects. The fluorescence lifetime ($\tau_f = 60$ ns) and yield ($Q_f = 0.11$) of the benzene-TCNE complex in organic rigid glasses at $T \le 100$ K correspond to $k_{CT}^{nr} \approx 1.6 \times 10^7$ s^{-1}, only slightly larger than that of exciplexes [22]. This rate increases rapidly with temperature in viscous organic solvents in the $T = 120$–150 K range and attains in the room-temperature fluid solutions the $10^{11} - 3 \times 10^{12}$ s^{-1} limits for different A + D pairs. It shows also a pronounced dependence on the solvent viscosity in the low as well as in the room-temperature range [22, 26–28].

The possible reason for these effects is the sloped intersection of diabatic CT and G surfaces (the avoided crossing of $|A\rangle$ and $|B\rangle$ surfaces) in the $R < R_{CT}^{eq}$ repulsive part of surfaces. The rate of the $CT \to G$ relaxation is sharply enhanced for the CT levels in the vicinity and above the crossing point so that the average relaxation rate is increased when these levels are thermally populated. The energy of the crossing point $U(R_c)$ is low in charge-transfer complexes (Fig. 6.3) and much higher in exciplexes. The thermal population of short-lived levels at room temperature is significant in the case of charge-transfer complexes but not in that of exciplexes. An argument in favour of this model, analogue of the inverted region of the Marcus theory, is an exponential increase of the decay rates with decreasing CT–G energy gap observed in room-temperature conditions [31–33].

The viscosity effect on the decay rates may be due to the dependence of the $U(R_c)$ energy on the mutual orientation of the complex components analogue to that observed for LE–CT transition (see Sect. 6.2.2). The CT–G decay rate depends, in this case, on that of its reorientation sensitive to the microviscosity of the solvent.

Direct evidence of the intersection of CT and G surfaces is the rapid ($k_{CT}^{nr} \approx 10^{12}$ s^{-1}) relaxation of jet-cooled iodine complexes with benzene [34–36], dioxane, diethylsulfide and acetone [37]. This process populates high vibrational levels of the ground electronic state of iodine, with energies exceeding the dissociation onset of I_2. The $CT \to G^{\#}$ relaxation yields thus free iodine atoms,

$$D^{+}..I_2^{-} \to D^{\#} + I_2^{\#} \to D^{\#} + I + I.$$

The case of iodine complexes is, however, particular because of the presence of low-lying repulsive states of I_2, which may play a role as intermediates of the relaxation process.

$LE \to CT$ Transitions

We will treat here the A+D systems composed of free A and D molecules and bichro-mophores A–B–D with a flexible or semirigid bridge (the A–B–D systems with rigid bridges merit a distinct discussion). We take as a model the same donor–acceptor pair A = anthracene (An) and D = dimethylaniline (DMA) in the A+D and A–B–D cases. These systems have been the object of the first pioneering studies of the electron transfer in solutions between A* and D species free [38] or bound by a -(CH$_2$)$_n$-($n = 3, 4$) flexible bridge [39].

The transition from the initially prepared $|LE\rangle$ state to the tightly bound contact complex in the CT state is a two-step process involving the electron jump, A*...D $\to$ A$^-$...D$^+$, followed by reorganization of the system configuration from the LE to the CT equilibrium geometry. The rate of the electron transfer, k_{ET}, may be determined by measurements of the decay of the A* $\to$ A LE fluorescence and of the A* $\to$ A** transient absorption of the neutral A* or of the rise of the absorption of the free A$^-$ and D$^+$ ions. The formation of the A$^-$D$^+$ CT state with the rate k_{CT}^{rise} may be monitored by detection of the rise (or displacement) of the $CT \to G$ emission. The sequence of the two processes depends on the nature of the A..D system.

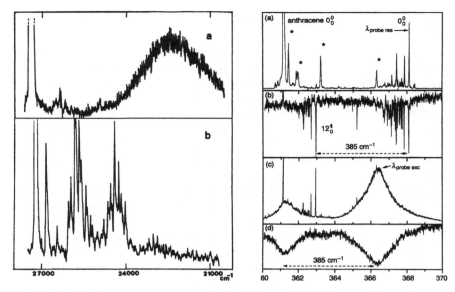

Fig. 6.7. At left, fluorescence spectra of the (a) E- and (b) R-forms of the anthracene-dimethylaniline complex. At right in (a) and (c), the fluorescence excitation spectra of both R and E species and in (b) and (d) the hole-burning spectra recorded at the frequencies indicated by arrows (from [41])

Jet-Cooled AD Complexes

The initial A..D distance in the LE state prepared by optical excitation $R_0 = R_G^{eq}$ and the equilibrium distances, R_{LE}^{eq} and R_{CT}^{eq}, are close to the sum of the van der Waals radii, $d_A + d_D$, but the energy and the $LE–CT$ coupling strength of an AD contact complex depend not only on R_{AD} but also on the angular coordinates. The proof of this dependence is the existence of isomeric forms of the AD complex differing by their spectra and dynamics.

The AD $\rightarrow$ AD* fluorescence excitation spectrum of the jet-cooled An-DMA complex corresponds to the slightly red-shifted $S_0 \rightarrow S_1$ band of anthracene. This spectrum is, however, composed of two systems of vibronic bands due, as shown by the hole-burning technique, to two isomeric forms, called, in the following, E- and R-isomers in view of their exciplex-like (E) and resonant (R) emission spectra (Fig. 6.7) [40, 41].

Upon the excitation of the E form, only the strongly red-shifted (the Stokes shift, $\Delta\omega_{St} \approx 5,000$ cm^{-1}) structureless broad ($\delta\omega \approx 2700$ cm^{-1}) band of the $CT \rightarrow G$ fluorescence is observed (Fig. 6.7 [left (a)]). The absence of the $LE \rightarrow G$ resonant emission indicates that the $LE \rightarrow CT$ transition (electron transfer) is much more rapid than the intrinsic decay of the LE state, $k_{ET} \gg k_{LE} \approx 5 \times 10^7$ s^{-1}. This rate is compatible with the widths ($\delta\omega \approx 50–150$ cm^{-1}) of vibronic bands in the spectrum of the E form (Fig. 6.7 [right (c), (d)]) corresponding to decay rates contained in the $2 \times 10^{12} - 2 \times 10^{13}$ s^{-1} limits. The anthracene-like vibrational structure of the excitation spectrum indicates that the initially excited state is a LE state perturbed in the vicinity of the $LE-CT$ intersection. The electron transfer is obviously the primary step followed by a small change of the system geometry.

The fluorescence-excitation spectrum of the R form of the An-DMA complex shows a fine structure of densely spaced narrow ($\delta\omega \approx 1$ cm^{-1}) lines assigned to combinations of internal modes of A with external modes of AD. Upon the excitation of the lowest ($E_{vib} < 50$ cm^{-1}) levels of the LE state, only the narrow-band $LE \rightarrow G$ fluorescence with the anthracene-like vibrational structure is emitted. Above this threshold, the emission spectrum is composed of the broad-band CT emission and of narrow LE emission bands. The intensity ratio I_{CT}/I_{LE} increases rapidly with E_{vib} so that only the CT emission is observed for $E_{vib} \geq 385$ cm^{-1}.

In the absence of time-resolved spectra, the transition rate from high levels may be only estimated from the quenching of the LE emission and linewidths in absorption spectrum as 10^8 s$^{-1} \ll k_{ET} \leq 10^{11}$ s^{-1}, smaller by one or two orders of magnitude than that of the E form. In spite of a close AD contact in the LE state, the $LE \rightarrow CT$ transition is significantly slower in the R configuration than in that of the E form.

The k_{ET} rates from high levels of the R form are of the same order of magnitude as the IVR rates in molecular complexes. One can suppose that the rate-limiting step is the vibrational energy redistribution among external modes allowing a random walk from the R to the E configuration followed by a rapid electron transfer. Unfortunately, the structures of R and E isomers are not known in spite of reliable calculations of the geometries and energies of the energy minima at the ground-state surface (cf. [40]).

A pronounced energy dependence of ET rates was reported for other complexes with R-type excitation spectra. For instance, the ET onset of 400 cm^{-1} and k_{ET} rates increasing from 3×10^7 to 10^9 s^{-1} for E_{vib} varying from 400 to 500 cm^{-1} are reported for the cyano-naphthalene + triethylamine system [42].

A close analogue of ET in jet-cooled AD complexes is the electron transfer from the donor solvent such as aniline or DMA to electronically excited solute such as rhodamine 6G, coumarine and oxazine-1, which may be considered as taking place within an AD_n cluster in fluid, room-temperature solutions [43–47]. The directly measured ET rates of $\sim 2 \times 10^{13}$ s^{-1} are of the same order of magnitude as those deduced from the bandwidths of E-isomer spectra and may be assigned to AD pairs with E-type structure. There must be a wide distribution of the cluster structures and the strongly non-exponential decay of the A^*D_n $|LE\rangle$ state was tentatively explained by the pronounced dependence of the electron-transfer rate on the configuration of the solvation shell [45, 47].

A–B–D Systems with Flexible and Semirigid B Bridges

In the initially prepared LE state of the jet-cooled 9-An-(CH$_2$)$_3$- p-DMA molecule (cf. Fig. 6.2) (analogue of the free An + DMA pair), the A and D groups are far apart. Its fluorescence excitation spectrum (nearly the same as that of the 9-alkyl anthracene) shows that anthracene is perturbed by the alkyl side chain but no A..D interaction can be detected. The spectrum is compatible with the *all-trans* configuration of the chain maintaining the distance $R_{DA} = R_0 \sim 5.5$ Å. Upon the excitation of vibronic levels

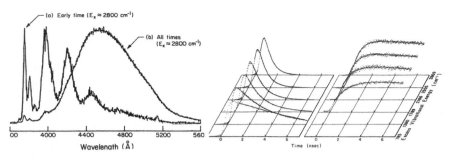

Fig. 6.8. Emission spectra and the time dependence of the A^* and $A^- D^+$ fluorescence intensity of the An-$(CH2)_3$-DMA jet-cooled molecule (from [15])

with $E_{vib} < 900$ cm^{-1} uniquely, the anthracene-like LE fluorescence is emitted, whereas, above this threshold, the emission spectrum is composed of narrow LE bands and broad-band CT emission, with the intensity ratio I_{CT}/I_{LE} increasing with the vibrational energy excess. The decay rate of the LE emission, k_{ET}, identical to k_{CT}^{rise} of the CT fluorescence, increases also with E_{vib} (Fig. 6.8) [15]. A similar energy dependence was observed in the case of other flexible spacers [48]. Such a time evolution was tentatively explained by assuming that an efficient $D \rightarrow A$ electron transfer does not occur at the initial A..D distance, $R_{AD} \approx R_0$, but only at $R_{AD} \leq R_{ET}$, where R_{ET} is the maximal distance for an efficient electron transfer. Such a structure may be attained by *interconversion* of the chain, which necessitates the vibrational energy exceeding the height of energy barriers between different chain configurations. This energy must be transferred from initially excited optically active modes of the anthracene ring to the bending modes of the chain [15]. The rate of the $LE \rightarrow CT$ electron transfer is thus determined by the probability of the $R_0 \rightarrow R_{ET}$ interconversion in the LE state dependent on the amplitude (energy) of chain vibrations and limited by the IVR rate. The 900 cm^{-1} threshold corresponds to the energy difference $U(R_{ET}) - U(R_0)$ or to the height of the energy barrier.

An alternative model is the *harpooning* mechanism [49, 50], assuming a long-range electron tunneling in the initial A–B–D configuration at $R_{AD} = R_0$ followed by a folding of the molecule driven by the $A^-..D^+$ Coulomb attraction [13, 49]. These two mechanisms of the $LE \rightarrow CT$ transition in a model A–B–D molecule are schematically represented in Fig. 6.9 [49]. They differ by the predicted time evolution of the initially excited A^*–B–D state:

- In the interconversion model, the rate-determining step is the transition from the ground-state equilibrium configuration with a large A..D distance R_0 to the partially folded $R_{AD} \leq R_{ET}$ configuration taking place at the LE surface. The next step consists of a $D \rightarrow A^*$ electron jump and a rapid evolution on the CT surface to a stable contact configuration. The quenching of LE emission and the appearance of the CT emission are quasi-simultaneous and delayed with respect to the initial excitation ($k_{LE}^q = k_{CT}^{rise}$).

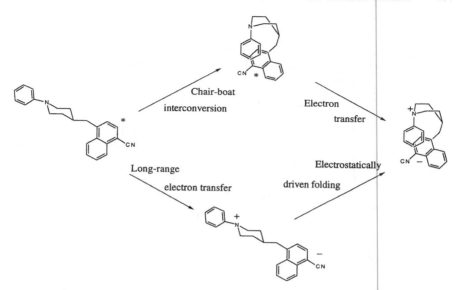

Fig. 6.9. Schematic representation of the interconversion and harpooning electron transfer mechanisms (from [49])

- In the harpooning model, the first step is a rapid formation of an extended CT state $A^- - B - D^+$, with $R_{AD} \approx R_0$, followed by a slower evolution to the stable compact configuration so that the appearance of the CT emission is delayed with respect to the LE emission decay ($k_{LE}^q > k_{CT}^{rise}$).

The kinetics of the An–B–DMA system correspond to the interconversion mechanism: the rates of the LE decay, k_{LE}, and of the CT rise, k_{CT}^{rise}, are identical and increase monotonically with the vibrational energy excess, i.e., with the amplitude of chain vibrations determining the probability of an A–D collision, with $R_{AD} \leq R_{ET}$.

On the other hand, the harpooning mechanism was evidenced already in the pioneering studies of the electron transfer between free An* and DMA in solutions [38]. The long-range $A_s^* + D_s \rightarrow A_s^- + D_s^+$ electron jump at the A*..D distance of ~ 7 Å was evidenced by the transient absorption of the A^- ion. The ions are stable in polar solvents, whereas, in weakly polar ones, the Coulomb attraction induces the $A_s^- .. D_s^+ \rightarrow (A^- D^+)_s$ reaction, giving rise to the fluorescent CT state.

The role of the harpooning mechanism was demonstrated in the case of a group of molecules composed of the aniline (donor) and cyano-naphthyl (acceptor) groups linked by the piperidine bridge (Fig. 6.2), with a high barrier between the chair and boat configurations. In room-temperature fluid solutions, the strongly red-shifted fluorescence emitted upon the excitation of the naphthyl group is that of the compact $A^- D^+$ complex, but in the viscous solvents, in which the folding is hindered, the emission spectrum contains also a higher frequency "(blue)" band assigned to the extended CT state, $A^- - B - D^+$. The blue emission decays while the red one builds up with the rate limited by the micro-viscosity of the solvent [51].

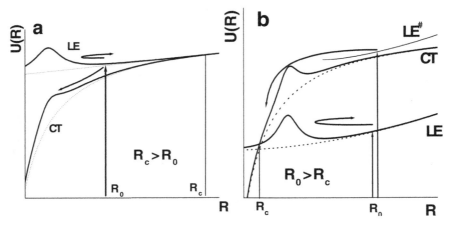

Fig. 6.10. The excitation and further evolution of the A-B-D system excited in (a)$R_0 < R_c$ and (b) $R_0 > R_c$ case

The respective roles of the harpooning and interconversion mechanisms depend on two factors: the $U_{CT}(R_0) - U_{LE}(R_0)$ energy difference at $R_0 \approx R_G^{eq}$, the initial A..D distance and the height of the energy barriers for formation of the contact complex. Its height depends only on the structure of the B bridge and is the same in the LE and CT states (Fig. 6.10), but on the steep CT surface, its effect is attenuated by the coulombic attraction. One can assume that, in view of the difference of the transition moments ($\mu_{LE} \gg \mu_{CT}$), the optical excitation prepares always the LE state. When $U_{CT}(R_0) - U_{LE}(R_0) < 0$, i.e., in the $R_0 < R_c$ case, the transition from LE to the extended CT is exoergic ($U_{LE}(R_0) > U_{CT}(R_0)$ and quasi-instantaneous and the further evolution on the strongly attractive CT surface is rapid (Fig. 6.10 (a)).

When $U_{CT}(R_0) - U_{LE}(R_0) > 0$ ($R_0 > R_c$), the $LE \rightarrow CT$ channel is closed in the vibrationally cold molecule. When the energy barrier is low, the system evolves on the shallow LE surface until it attains the R_c crossing point, where the $LE \rightarrow CT$ transition takes place, but this evolution is prevented by the barrier (Fig. 6.10 (b)). The barrier may be crossed when the LE state is vibrationally hot, with the vibrational energy excess E_{vib} exceeding the energy difference $U_{CT}(R_0) - U_{LE}(R_0)$. The $LE^{\#} \rightarrow CT$ transition takes place and, once on the CT surface, the system may attain the equilibrium configuration of the contact complex. This channel may be open even for E_{vib} smaller than the barrier height, but its efficiency decreases with increasing energy difference, $U_{CT}(R_0) - U_{LE}(R_0)$. The interconversion mechanism is then more efficient than harpooning, as in the case of An-B-DMA complex.

The formation of the compact CT states in the molecules with the piperidine bridge involves the harpooning mechanism. The chair-to-boat isomerization barrier of 3,500 cm^{-1} is too high to allow such a transition on the LE surface. This channel is open, with the energy threshold lower than the barrier height ($\sim$1700 cm^{-1}), reduced to 750 cm^{-1} upon the substitution of the phenyl ring by the -OCH$_3$ group, which has no effect on the energy barrier but decreases the ionization potential of the donor group. The threshold energy does not correspond to the barrier height, but to the energy $\Delta U(R_0)$ necessary for transition to the extended CT state of $A^-..D^+$ at $R_{DA} = R_0$; its further evolution on the strongly attractive CT surface is practically barrierless [13, 49].

A–B–D Systems with Rigid B Bridges

A and D groups bound by a rigid norbornyl bridge are maintained at a constant distance, R_n, dependent on the number n of the bridge C-C σ bonds and equal roughly $R_n \approx 1.35$ n$\overset{\circ}{A}$ for $n = 4$–12.

In the fixed geometry, only the extended CT states exist, the formation of compact CT states being excluded. The time-resolved microwave measurements show that dipole moments of CT states $\mu_{CT} \approx eR_n$ correspond to the complete charge separation [14, 53]. The energies of the A–B–D* LE states are practically n-independent, whereas those of the extended CT states stabilized by the coulombic term $C_n \approx -e^2/R_n$, increase rapidly with n.

The interest of systems with rigid bridges consists in the possibility of determining the dependence of the rates of the long-range electron transfer $LE \to CT$ and $CT \to G$ processes on the D..A distance. These rates are closely related to the overlap between donor and acceptor orbitals, S_{DA}, and decrease exponentially with R,

$$k_{ET}(R) = V_0 e^{-\beta R}.$$

This decrease is slower in the case of bound A-B-D as compared with that of free A + D pairs. The reason for this difference is the hyperconjugation of π-electron orbitals of A and D with σ orbitals of the bridge CH_2 groups. The through-bond interaction along the B wire in A-B-D increases the efficiency of the electron transmission with respect to that of the through-space interaction in an A...D pair in vacuo. [10, 52].

Let us take as a model the system composed of the di-methoxy- naphthalene donor and dicyano-ethylene acceptor separated by an $n = 4$–13 σ norbornyl bonds. The $LE \to CT$ transfer takes place in the compounds with $n \le 8$, including $n = 8$, with nearly isoenergetic CT and LE states, whereas the CT state of the $n = 10$ compound with $U_{CT} > U_{LE}$ is not populated. The rate constants $k_{LE \to CT}$ decrease with n, but even for $n = 8$ with $R_n \approx 10.5$ $\overset{\circ}{A}$ distance, the transfer is rapid, $k_{LE \to CT} = 5 \times 10^{11}$ s^{-1}. This value is surprising in view of the $R_{ET} < 5.5$ $\overset{\circ}{A}$ in the An–B–DMA case, but it must be kept in mind that the dicyano-ethylene group is a much stronger electron acceptor than anthryl and that we compare an isolated, jet-cooled molecule with a room-temperature solution. The rate of the $CT \to G$ charge recombination is much smaller (probably because of a reduced value of the overlap integral S_{DA}) and decreases with n until $n = 8$. At $n = 8$, the CT state quasi-isoenergetic with the LE state decays by the reverse $CT \to LE$ relaxation, as evidenced by the delayed LE fluorescence [53].

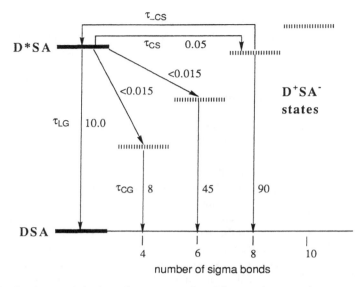

Fig. 6.11. The characteristic times (in nanoseconds) of direct and reverse electron transfer rates as a function of the number of intermediate bonds in a rigid norbornyl structure (from [53])

6.3 Proton/Hydrogen-Atom Transfer

6.3.1 General Notions

As previously mentioned (Sect. 6.1), the processes induced by electronic excitation of a large number of molecular systems with intramolecular DH..A or intermolecular A-DH..C hydrogen bonds are described in the diabatic basis as proton or hydrogen-atom transfer.

We are interested in molecular systems in which the DH..A is the stable structure in the ground state whereas (D..HA)* is stable in the excited electronic state. The light-pulse excitation prepares, thus, an unstable Franck–Condon configuration (DH..A)*. Its further isomerization, (DH..A)* → (D..HA)*, is monitored by the time-resolved detection of the spontaneous or stimulated emission and of the transient absorption. It is possible to realize a similar experiment with a ground-state molecule prepared in an unstable configuration D..HA, e.g., by the stimulated emission from (D..HA)* state, but the analysis of data is more difficult in the presence of a large excess of unexcited DH..A molecules.

We will discuss separately the problems of the intra- and intermolecular H^+/H transfer processes.

6.3.2 Intramolecular Hydrogen Transfer

Enol-Keto Isomerism

The term intramolecular proton transfer is used for isomerization processes much

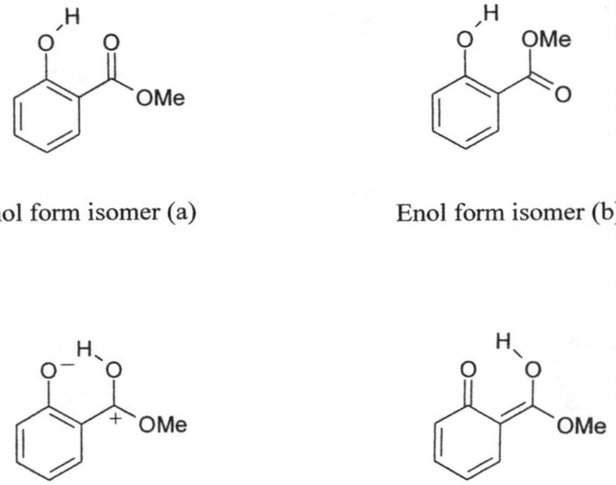

Enol form isomer (a)   Enol form isomer (b)

Zwitterion (c)   Keto form (d)

Fig. 6.12. Structures of isomeric forms of methyl salicylate (MS)

more complex than a simple displacement of a hydrogen atom. The classical case is that of the methyl salicylate (MS), in which the proton affinity of the carboxy group and proton-donor properties of the -OH group are increased by the electronic excitation [54, 55] (Fig. 6.12). The blue fluorescence spectrum of the isomer (a) shows an anomalously large Stokes shift of $\sim 7,000\ \text{cm}^{-1}$, whereas that of the form (b) is normal ($\Delta\omega_{St} \approx 2,500\ \text{cm}^{-1}$). This behaviour may be explained by the proton transfer in the C=O..H-O-C hydrogen bond with formation of a zwitter ion, but the studies of the solvent effects in solutions [55] show that the fluorescent form is stabilized in apolar solvents, which indicates a dipole moment smaller than that of the (a*), electronically excited isomer (a). It must be thus assigned not to the zwitter ion (c) but to the neutral keto form (d). Its geometry differs from that of (a) not only by the position of the H atom but also by other parameters, the most striking being transformation of a single C-O bond of the phenol group into a double C=O bond. This transformation was directly evidenced in the case of another model molecule with an intramolecular O-H..N bond, HBT (Fig. 6.13). The transient infrared absorption spectrum of HBT excited to the S_1 state contains the $\omega \approx 1,530\ \text{cm}^{-1}$ band characteristic for aromatic ketones, absent in its ground-state absorption spectrum [57].

The fluorescence and fluorescence-excitation spectra of methyl salicylate in rare-gas matrices [56] and in supersonic expansions [58, 59] confirm the assignment of the fluorescence with a large Stokes shift to the keto form. The same conclusion is valid for a large group of molecules with intramolecular O-H..O and O-H..N bonds. We will limit our discussion to two groups of model molecules with a very different electronic-charge distribution in their ground and excited $\pi\pi^*$ states (Fig. 6.13):

- The benzene and naphthalene derivatives with the hydroxy and carboxyl groups in ortho position as in the MS molecule and the enol (E) ground-state structure

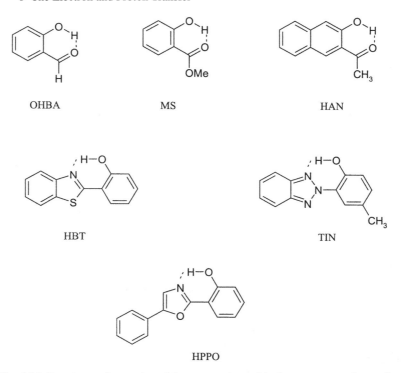

Fig. 6.13. Structures of several model compounds used in the proton transfer studies

OHBA, HAN. Their broadband emission spectra, Stokes shifted to the red by 5,000-6,000 cm^{-1}, are assigned to the keto form.

• The oxazole and thio-oxazole compounds with the C-O-H..N bridges in the ground state and the fluorescence of the C=O..H-N forms still more ($\Delta\omega_{St} \geq 10,000$ cm^{-1}) shifted to the red (HBT, TIN, HPPO).

Potential-Energy Surfaces and Dynamics

The ground-state, (a) and (d), and excited-state, (a*) and (d*), enol (E) and keto (K) isomeric forms are *diabatic states* of the D..H..A molecule and the intramolecular proton transfer corresponds to the a* → d* (E* → K*) and d → a (K → E) isomerization processes. These processes are, however, better described in terms of *adiabatic surfaces A and B*.

The multidimensional potential energy surface of the D-H..A system is usually approximated by a function of two coordinates, the length of the D-H bond, r_{DH}, and the D..A distance, R_{DA}, the other ones being considered as the heat bath. The R_{DA} variable represents, in fact, a combination of all relevant heavy-atom coordinates. Its reduction to a one-dimensional potential curve, by assuming a complete separation of the rapid proton motion from other coordinates, is useful for illustration of the

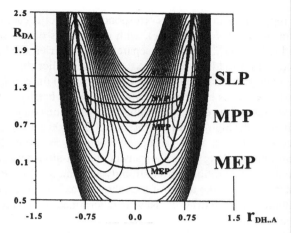

Fig. 6.14. The main proton-transfer paths: (SLP, straight-line path; MEP, minimum-energy path and MPP, maximum probability path) at the symmetric double-minimum surface (from [60])

proton-transfer process in terms of a set of potential curves, $U_n(r_{DH})$, drawn for each electronic state, $|n\rangle$, at a constant R_{DA} distance (Fig. 6.15). The one-dimensional representation is, however, not sufficient for description of finer details of processes involving the strong coupling of the proton movement to that of the heavy nuclei.

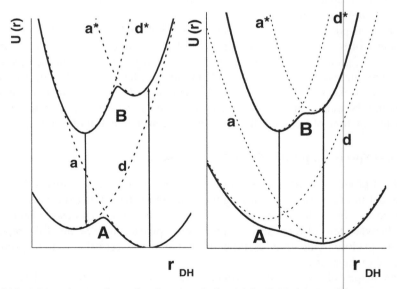

Fig. 6.15. Schematic one-dimensional representation of the D-H..A potential energy surface with and without potential-energy barriers

The surface represented in Fig. 6.14 corresponds to a specific case of a symmetric double-minimum potential, $U(r_{DH}, R_{DA})$, with coordinates of the energy minima, $R_{DA}^{eq'} = R_{DA}^{eq}$ and $r_{DH}^{eq'} = -r_{DH}^{eq}$. They are separated by a barrier, the height of which depends on R_{DA}, with a saddle point at R_{DA} distance smaller than the equilibrium distance, R_{DA}^{eq}, and may be even reduced to zero in this part of the configuration space. The proton transfer proceeds either by tunneling across the barrier or by turning it.

In the close vicinity of the energy minimum, the vibrations may be considered as harmonic local modes, the high-frequency ($\omega_r = 2,000$–$3,500$ cm^{-1}) proton vibration along the r_{DH} coordinate and a low-frequency mode ($\omega_R = 100$ to 200 cm^{-1}), the movement of heavy nuclei along R_{DA}. The transfer of the H-atom between two minima involves large-amplitude vibrations on the strongly anharmonic $U(r, R)$ surface, with a frequency ω dependent on the angle of the vibration axis with respect to r and R axes. The probability of the barrier crossing is determined by ω and by the barrier height ΔU in the crossing point,

$$k = \omega e^{-\alpha \Delta U}.$$

One can presume that the most probable path (MPP) from one minimum to the other will be intermediate between the straight-line paths (SLP) and minimum-energy path (MEP), indicated in Fig. 6.14. The choice of the best path depends on the details of the surface. The direct SLP is handicapped by the large barrier height ΔU (R_{DA}^{eq}) in spite of the large pre-exponential factor $\omega \approx \omega_r$. The probability of the barrier crossing on the minimum-energy path (MEP) through the saddle point is higher, but this path necessitates large displacements of heavy A and D groups with the low frequency $\omega \approx \omega_R$ [60].

The short-pulse $A \rightarrow B$ excitation of the vibrationally cold A..HD is described as creation of the vibrational wave packet, $\mu F |0\rangle$, centered in the (r_0, R_0) Franck–Condon configuration. This point does not necessarily coincide with the secondary energy minimum of the B-state surface but is not very far from it. The time evolution of the wave packet from the initial (Franck–Condon) to the final (stable) configuration cannot thus be treated in terms of kinetics but as its travel on the (r, R) surface. The characteristic time of this evolution must not be considered as a rate constant.

Ultrafast Spectroscopy of H-Transfer Processes

Important progress in the understanding of the H-transfer mechanisms is due to the recent real-time experiments applying the pump-probe techniques (cf. Sect. 2) at the femtosecond time scale. The characteristic times of the transfer processes vary in wide limits between tens of femtoseconds and several picoseconds. Some of the reported rates are so close to the time-resolution limit that their exact values will be probably corrected in the near future.

The shapes of the potential surfaces (energy differences between DH..A and D..HA forms) are sensitive to the solvent, but the ultrafast dynamics on these surfaces is practically not affected by the environmental effects because the solvent movements are slow as compared with intramolecular processes. The rates measured for the molecules, collision-free and dissolved in inert solvents, are thus nearly the same.

Enol-Keto $E^* \rightarrow K^*$ Transitions

In both types of compounds, the $E^* \rightarrow K^*$ relaxation is so rapid that the fluorescence of the enol form is completely quenched and the rate of the transition is estimated only from the rise time of emission (or transient absorption) of the K^* form. For instance, when the E^* state of HBT with $E_{vib} \approx 2{,}000$ cm^{-1} is initially prepared, the rise time of the K^* isomer $t_{rise} \leq 60$ fs is close to the time-resolution limit. The same rate is reported for both -OH and -OD species in the room-temperature gas and in solution [61]. This time scale corresponds to a few classical periods of the O-H and O-D stretching and could be assigned to the direct $E^* \rightarrow K^*$ transition involving the straight-line SLP path and tunneling through the energy barrier. This mechanism implies, however, a large deuterium effect due to the mass dependence of the tunneling probability and of the ω_r pre-exponential factor. In the absence of any mass effect, the SLP trace of the $E^* \rightarrow K^*$ transition seems to be excluded. The more probable mechanism is a travel of the wave packet through the barrierless part of the surface, with transit at the time scale of low-frequency vibrations of the D..A oscillator.

The energy barriers in some of the O-H..O systems are, however, nonnegligible. The linewidths in the $(E \rightarrow E^*)$ fluorescence excitation spectra of the jet-cooled molecules, such as HAN and HBT derivatives (HPPO), correspond to the $E^* \rightarrow K^*$ characteristic times t_{rise} of the order of 1 ps with a pronounced deuterium effect in the case of HPPO [62]. This time is reduced to $\tau \leq 100$ fs for the gas-phase HAN molecule, with E_{vib} of the order of 2,000 cm^{-1} [63], which suggests that the barrier height is relatively low.

Coherent Effects

The rapid rise of the stimulated emission and transient absorption of K^* states recorded upon the short-pulse excitation of the E^* states is modulated by oscillations with the 50 to few hundred femtosecond periods (Fig. 6.16). The Fourier-transform spectrum is composed of a few bands corresponding to frequencies (or differences between frequencies) assigned to vibrational modes of the keto form: $\sim$250 and $\sim$470 cm^{-1} in TIN [64], 118, 254 and 529 cm^{-1} in HBT [65] and 50–60 cm^{-1} in jet-cooled OHBA derivatives [66].

The initial vibrational state of the K^* form corresponds obviously to a coherent superposition of a few vibrational modes. This superposition is not created by the optical excitation $E \rightarrow E^*$ but by a rapid $E^* \rightarrow K^*$ jump, which prepares the K^* state with a large vibrational energy excess shared between a few active modes. This process may be visualized as the oscillations of the wave packet penetrating into the deep energy well of the keto form (Fig. 6.17).

The oscillations are damped at a few picosecond time scale, which suggests the energy flow from the active modes to other modes of the molecule and of its solvation shell. The time scale for the decay of the oscillation amplitudes is not significantly shorter in solutions than in the supersonic expansion; one can thus consider that the damping rate is controlled by the intramolecular process of vibrational redistribution (IVR).

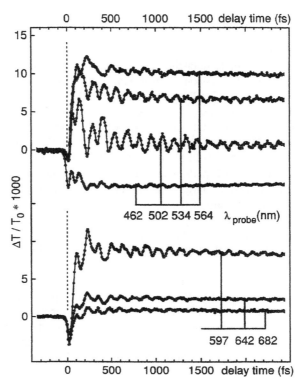

Fig. 6.16. Oscillating behaviour of the transient absorption/stimulated emission spectrum of the electronically excited HBT molecule (from [65])

Similar quantum beats in the CT states induced by an ultrafast electron transfer have been observed in the spectra of charge transfer complexes [26] and of electron-acceptor solutes in electron-donor solvents [46, 47]. They seem also to be due to an ultrafast transition to the new state with a large vibrational energy excess concentrated in a few vibrational modes. This analogy between the ultrafast electron-transfer and proton-transfer processes is interesting but not really elucidated.

Decay of Electronically Excited Keto K^* Isomers

The K^* excited species decay with a rate varying in extremely wide limits: in TIN, τ_{K^*} is equal to 150 fs and is independent of the excess energy in the initial E^*-form in the 0-5,000 cm^{-1} limit [65] but is of the order of 3–6 ps in OHBA [66] and of $\sim$300 ps in HBT [68]. Those of jet-cooled MS [58] and HAN [69] at $E_{\text{vib}} \approx 0$ are still longer (of the order of 10 ns) but reduced to $\sim$100 ps for the energy excess of the order of 2,000 cm^{-1}.

The possible decay channels are a rapid intersystem crossing (ISC) characteristic for aromatic ketones, a transition to the $^1n\pi^*$ state [70] or a direct $K^* \rightarrow K^{\#}$ internal conversion (IC). The competition between these channels was discussed in the case

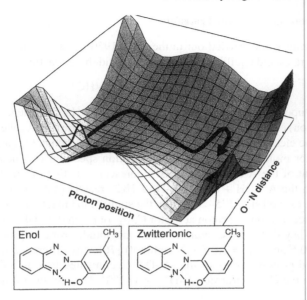

Fig. 6.17. Schematic representation of the trajectory of the wave packet on the excited-state surface (from [67])

of hydroxy-flavones [71–73]. The information about the competition between the IC and ISC channels may be deduced from the recovery rate of the $E \rightarrow E^*$ absorption bleached by the pump pulse; its rapid recovery is expected when IC is predominant: the vibrationally excited ground-state $K^{\#}$ decays by the $K^{\#} \rightarrow E^{\#}$ transition. The recovery of the E-state population will be delayed when the long-lived triplet state is populated via the ISC channel. In TIN, the $K \rightarrow E$ proton back transfer, with a time constant of 600 fs, is observed [74], which indicates an efficient IC path. No data for other systems are available.

A large difference between the rates of the $K^* \rightarrow K^{\#}$ internal conversion may be related to the shapes of the excited and ground-state surfaces $U_B(r, R)$ and $U_A(r, R)$ in the vicinity of the energy minimum of the K^* form. The energy gap between $U_B(r, R)$ and $U_A(r, R)$ surfaces is reduced in this region, as evidenced by a large Stokes shift of the K^*-fluorescence, and the $U_A(r, R)$ surface is strongly repulsive in this region (Fig. 6.15). The relaxation rate is sensitive to the energy of the crossing between diabatic surfaces (curves d* and a of Fig. 6.15). A correlation between the efficiency of internal conversion and the Stokes shift is expected. The same model of the sloped intersection of the repulsive parts of A and B potential surfaces has been used previously in this work for description of the rapid relaxation of the CT state of charge-transfer complexes (see Sect. 6.2.2).

6.3.3 Intermolecular Proton Transfer

In the pioneering works carried out in the early 1950s, it was shown that, in water solutions, the rates and equilibrium constants of acid–base reactions,

$$A - DH + C \leftrightarrow A - D^- + HC^+,$$

are displaced upon the electronic excitation of the A–DH acid (phenol, naphthol or anilinium cation) composed of an electron-acceptor aromatic ring A (phenyl, naphthyl etc.) and of the electron donor D (O or N atom). In this scheme, C is a weakly basic proton-acceptor molecule or anion. The absorption spectrum in a neutral $p_H \approx 7$ solution is that of the A-DH acid but the fluorescence is that of the A*–D$^-$ anion resulting from the A*–DH + C → A*–D$^-$+ HC$^+$ reaction taking place during the lifetime of the excited state [75–76]. From the study of the reaction rate as a function of p_H and concentration of proton acceptor C, p_K^* of excited A–DH molecules were determined. They differ by several orders of magnitude from its ground state, p_K (e.g., $p_K = 10.5$ and $p_K^* = 3.0$ for 2-naphthol) [76]; this difference was explained by the reduced proton affinity of the $(A–D^-)^*$ conjugated base in its excited state resulting from the charge transfer $A–D^- \rightarrow (A^{-\delta})^*–D^{\delta-1}$ in the excited electronic state [77].

The free $A–D_s^-$ and CH_s^+ ions are stable only in polar hydroxylic solvents (water, alcohols), stabilized by strong solvent-to-solute hydrogen bonds, $A–D^-$..H–OR and CH^+..OR. In weakly polar solvents, the electronic excitation of an A–DH..C complex gives rise to a strongly polar $(A–D^{-\delta})^*$..HC$^{+\delta}$ structure, as evidenced by its emission spectrum similar to that of the $(A–D^{-\delta})^*$ anion [78]. This structure seems to be better described in terms of the adiabatic excited-state surface,

$$|B(R, r)\rangle = \alpha(R)|a(R, r)\rangle + \beta(R)|b(R, r)\rangle,$$

where $a(R, r)$ and $b(R, r)$ are diabatic neutral, A–DH..C, and ionic, A–D$^-$..HC$^+$, states.

The values of $\alpha(R)$ and $\beta(R)$ coefficients, strongly varying with R, depend not only on the properties of A-DH and C but also on dielectric properties of the medium. So, the ion-pair structure of an isolated hydrogen-bonded complex is unstable in excited as well as in the ground electronic state. The structures of collision-free 1:1 complexes of strong acid and bases are neutral. The rotational spectra indicate the A-D..H..C structures with the proton shared between A-D and C only in the complexes of the strongest acids and bases, such as $(CH_3)_3N.HX$ complexes (where X = F, Cl, Br, I), whereas that of $H_3N.HX$ and pyridine.HX corresponds to a pair of neutrals [79, 80]. To our best knowledge, the only reported case of the proton transfer in an isolated 1:1 complex of an aromatic acid is that of the o-cyanophenol-trimethylamine complex [81].

The studies of the intermolecular hydrogen bonds in electronically excited systems are thus practically limited to condensed phases in which the intra- and intermolecular effects cannot be easily separated and the structures of AD–H..C are not well defined in view of the wide distribution of the configurations of the complex and of its solvation shell.

The influence of the environment on the proton-transfer process was demonstrated in the studies of jet-cooled clusters A–DH.C_n, in which the A–DH acid (phenol, 1- and 2-naphthol etc.) is solvated with the base C: H_2O, NH_3, $(CH_3)_2NH$. The proton transfer induced by the electronic excitation of AD-H is evidenced by fluorescence spectra and ionization potentials of the excited system. A critical review of these works may be found in [82, 83]. The proton transfer does not occur in 1:1 complexes but only in clusters with the number of ligands C $n \geq n_{min}$. This n_{min} onset depends on the basicity of C: in the case of 1-naphthol, $n_{min} = 2$ for $(CH_3)_2NH$ and 4 or 5 for NH_3 [82]. In still larger ammonia clusters, the proton is transferred already in the ground-state complex [83]. Similar effects were observed for AD–H.$(NH_3)_n$ clusters of phenol [84] and naphthols [85] in rare-gas matrices with the n_{min} threshold different in the S_1, T_1 and S_0 states differing by their acidities. The proton-transfer processes in excited $\pi\pi^*$ states are in several cases complicated by the presence of the dissociative $\pi\sigma^*$ states, which decay nonradiatively with formation of neutral (radical) products, $(A-D)^*H..C \rightarrow (A-D)^* \cdot +\cdot HC$ [83]. The same channel is probably responsible for the quenching of fluorescence of electronically excited aromatic enols by aromatic amines and azines, surprising in view of a high fluorescence yield of enol complexes with aliphatic amines [86, 87].

There is no reason to suppose that the mechanisms of the proton transfer in intermolecular and intramolecular bonds are different. One can thus expect that the A-DH..C $\rightarrow$ A–D^-..HC^+ transition will occur at the same subpicosecond time scale as in the DH..A systems. Surprisingly, much longer times, of 50–100 ps, were reported for A-DH/C_n clusters in the gas phase [88] and in rare-gas matrices [85]. These times seem to be limited not by the proton transfer rate but by that of processes involving reorientation of molecules or of their active groups preceding or following the displacement of the proton. One can suppose that the barrier for the proton transfer so strongly depends on the geometry of the complex that the proton jump occurs only after reorientation of the complex to a convenient configuration. The overall rate is limited by that of the hindered rotation of the complex components. The strongly nonexponential character of the A–DH..C $\rightarrow$ AD$^-$..HC^+ process with a fast (within 150 fs) step corresponding probably to the proton jump [89, 90] may be considered as an argument of favor of this picture.

6.4 Final Remarks

The electron-transfer and proton (hydrogen atom) transfer processes are well de- scribed in the framework of the theory of radiationless transitions provided that the complex shapes of potential-energy surfaces are taken into account. For the electron- transfer phenomena, the treatment in terms of diabatic ionic and neutral states is convenient, whereas the proton transfer is better described in terms of strongly an- harmonic adiabatic surfaces.

As could be expected, the transitions involving the light particles are ultrafast if their rate is not limited by the processes necessitating large displacements of heavier particles. So, the electron transfer in an exciplex is rapid, but the rate of the overall

$A..D \rightarrow A^-D^+$ process is determined either by reorientation of the $A..D$ system to a configuration allowing the electron jump or by the evolution of the $A^-..D^+$ ion pair created by electron transfer to the stable structure of a tight contact complex. In a similar way, the measured proton-transfer time in the $A..H..D$ system is in fact that of displacement of A and D to a configuration convenient for the proton jump.

In condensed phases, the situation is still complicated by the finite relaxation times of fluid or rigid solvents. They do not influence directly the elementary transfer step but may hinder the intramolecular processes involving heavy particles. The electron and proton transfer itself seems to be so rapid (several tens femtoseconds) that it is practically decoupled from a slow movements of solvent molecules. The transition rates in solution, are nearly the same as in the case of isolated molecules, and persistence of coherent effects (quantum beats) in condensed phases may be explained by the weakness of this coupling.

In view of the progress in the femtosecond experiments, one can expect that more precise measurements of the relaxation rates will allow better characterizing of the relation between the transition times and structure of molecules and/or their environments.

7

Postface

The essential conclusion of this review is that the basic ideas of the theory survived in a good shape the flood of new experimental results. Its validity was corroborated by observation of such phenomena as the quantum-beat modulation of fluorescence decays. The progress consisted rather in a better adaptation of its general notions to different types of molecular systems.

1. A wide variety of relaxation processes cannot be correctly described in terms of simple schemes of zero-order $|n\rangle$ states in which the bright state, $|s\rangle$, is coupled to a single dark state, $|\ell\rangle$, or to a set of ℓ states characterized by an average coupling constant, $\langle V_{s\ell}\rangle$, and an averaged level spacing, $\langle \Delta E_{\ell\ell'}\rangle$, or level density, $\rho = 1/\langle \Delta E_{\ell\ell'}\rangle$.

In the major part of molecular systems, the level-coupling pattern is more complex which implies the *sequential character* of the time evolution of the initially prepared excited state. The coupling between the s and ℓ levels varies in such wide limits that, for a major part of them, one can assume that $V_{s\ell} = 0$. The bright, $|s\rangle$, state is efficiently coupled only to a small fraction of dark, $|\ell\rangle$, levels, which are coupled in turn to the remaining part of the dark manifold, which may contain a continuum or quasi-continuum $\{m\}$. This hierarchy is accounted for by a simple model assuming the s-ℓ-$\{m\}$ *sequential coupling* but more adequate is the *tier model* of a sequential coupling between several tiers of states forming the s-ℓ- $\ell'\cdots$-$\{m\}$ chains. In this picture, the notions of the average coupling constant and of the overall level density are meaningless. The time evolution of the initially prepared state is a *sequential* process in which the time scale of each s-ℓ or ℓ-ℓ' step is determined by its own parameters: the coupling constant, $V_{s\ell}$ ($V_{\ell\ell'}$), and the level spacing, $\Delta E_{s\ell}$ ($\Delta E_{\ell\ell'}$). The overall rate of the whole process is limited by that of its slowest step, as in the case of chain reactions in chemical kinetics.

Because the role of the overall level spacing (level density) is limited, the time scale of relaxation is determined mostly by widely varying values of the coupling constants.

The propensity rules governing the dependence of the coupling matrix elements, $\langle s|V|\ell\rangle$, on the properties of the $|s\rangle$ and $|\ell\rangle$ states take particular forms for different relaxation processes, monomolecular or induced by external perturbations, but correspond to different aspects of the general *momentum gap law* valid in the limits of validity of the Born–Oppenheimer approximation. This law states that, among different relaxation channels, those involving the smallest changes of linear (k) and angular (J, K, M) momenta and of vibrational quantum numbers (v) are favoured.

So, in the case of the intramolecular vibrational redistribution, the decrease of $V_{s\ell}$ with the increasing difference of vibrational quantum numbers, $\Delta v_{\text{tot}} = \sum_i |v_i^s - v_i^\ell|$, is due to the selection rules for anharmonic and Coriolis coupling. The same Δv_{tot} dependence of the vibronic and spin-orbit coupling between vibronic levels of different electronic states of an isolated molecule is a consequence of the Franck–Condon rule, which implies a rapid decrease of the $\langle \chi_v | \chi_{v'} \rangle$ overlap integral with increasing Δv_{tot} value. In the case of the coupling of the molecular level system with a continuum of energy (dissociation, medium-induced relaxation), the propensity to small changes of molecular parameters is completed by a propensity to small changes of the linear momentum, Δk, i.e., of the translational energy, $\Delta E \sim |k^2 - k'^2|$, transferred to (or from) the continuum. In view of the relations $\Delta E_{\text{vib}} \sim \Delta v$ and $\Delta E_{\text{tr}} \sim |k^2 - k'^2|$, the momentum-gap law is often called the *energy-gap law*.

It must be kept in mind that the validity of gap laws is limited to the systems in which the Born–Oppenheimer approximation is valid. The electronic relaxation or predissociation rates obey the gap laws when the coupling (vibronic or spin-orbit) of A and B electronic states. V_{AB}^{el}. is small as compared with the energy distance between adiabatic surfaces $U_A(Q)$ and $U_B(Q)$ in the whole relevant range of configuration Q. Therefore, the gap laws break down when $V_{AB}^{\text{el}} \approx \Delta U_{AB}(Q)$ in the vicinity of the surface intersection ($Q \approx Q_c$). On the other hand, when the rate of a sequential process is determined by that of its slowest step, one can observe an apparent breakdown of the gap law applied to the overall process and not to its rate determining step. Such a breakdown may be considered as a sign of the sequential character of a process. At last, in the processes involving an energy transfer, ΔE, to its heat bath, the efficiency of a relaxation channel depends on two factors, Δv_{tot} and ΔE_{tr}. The most efficient channel may then correspond to a compromise between these two propensities.

2. The relation between the rates of individual steps of the same sequential process occurring in different molecular systems depends on their level-coupling patterns. A good example is the vibrational predissociation of polyatomic molecules and complexes involving vibrational redistribution as the first step. The classical theories of unimolecular reactions (such as RRKM), assuming an instantaneous dissipative vibrational redistribution followed by a slow dissociation ($k_{\text{IVR}} \gg k_{\text{diss}}$), work well for large molecules in solutions. The $k_{\text{IVR}}/k_{\text{diss}}$ ratio is inverted in small molecular systems in which the slow, rate-determining step is the vibrational-energy redistribution followed by a rapid dissociation. The dissociation mechanism has an impact on the energy distribution in the dissociation products and on their reactivity.

3. One of the key problems of the theory is the transition from the small-molecule system of discrete, nonoverlapping levels to the large molecule, in which the ℓ manifold (or a part of it) may be approximated by a dissipative continuum. This corresponds, in the time domain, to the transition from the coherent, periodic evolution of a discrete-level system to the aperiodic irreversible decay. We consider that the ℓ-level set behaves as a continuum when the level widths exceed their spacing, i.e., when $\gamma_\ell / \langle \Delta E_{\ell\ell'} \rangle = \gamma_\ell \rho_\ell \geq 1$. This condition corresponds, in the time domain, to

the decay time of ℓ levels shorter than the recurrence time in the discrete s-ℓ system so that all recurrences are quenched.

The evolution from the small- to the large-molecule case may be thus induced either by an increase of the density, ρ_{eff}, of levels effectively coupled to $|s\rangle$ or by an increase of their widths, γ_ℓ. We believe that a major part of real systems is described better in terms of increased widths of the small-molecule levels induced by their coupling to the continuum or quasi-continuum. When the size of the system is increased by substitution or complex formation, the dense set of substituent or ligand levels behaves as a quasi-continuum $\{m\}$ responsible for broadening of ℓ levels. In the same way, the coupling of discrete ℓ levels to the continuum of the kinetic energy by gas-phase collisions or solvent–solute (host–guest) interactions implies the increase of their widths. Both mechanisms may be described in terms of the sequential-coupling or the tier model.

The important question from the practical point of view is to what extent the coupling parameters and the rates of processes occurring in the small-molecule system are modified by the heat bath involving the substitution and solvent effects. The answer will be different for different types of processes. In view of the extremely weak coupling of electronic spin with molecular movements, the rates of the singlet-triplet transitions determined by the spin-orbit coupling strength is nearly solvent independent in the absence of specific heavy-atom effects. The ultrafast transitions taking place at the femtosecond time scale, such as the electronic relaxation via conical intersections, are also unperturbed because the orientational relaxation of molecules forming the solvation shell is too slow to follow the intramolecular process. In contrast, the vibrational redistribution (IVR) rate is significantly perturbed in condensed phases because the coupling strength of internal-to-external vibrational modes and between different internal modes is of the same order of magnitude, implying similar time scales of intrinsic IVR and of the vibrational relaxation induced by the environment of the molecule.

4. In spite of rapid progress in the experimental studies of relaxation phenomena, wide fields remain still almost unexplored. The attention was focused on the closed-shell molecules in their ground and the lowest excited states, whereas the knowledge of odd-electron systems (radicals and ions) is still fragmentary. The high time- and energy-resolution was attained quite recently and the easily accessible spectral domain was limited to the narrow frequency range between near infrared and near ultraviolet.

The application of femtosecond techniques opens new, unexplored perspectives. The actual studies of the ultrafast internal conversion of higher singlet states of molecules may be considered as a first step in this field. The theoretical treatment of dynamics at the surface intersection needs further developments The successful real-time investigation of the intramolecular electron- and proton-transfer may be extended to a large class of ultrafast isomerization processes.

The properties of the excited molecular system are determined not only by the frequency, ω; width, $\delta\omega_{\text{coh}}$, and intensity distribution, $I(\omega)$, of the exciting pulse but also by the phase relations and the pulse shape, the parameters that were practically disregarded in molecular physics. They seem to play an important role in the case of femtosecond light pulses with the $I(\omega)$ intensity distribution varying within the pulse duration. It is possible to realize in this way a selective multiphoton excitation of individual energy levels above the dissociation and ionization thresholds leading to the control of photochemical reactions. The dynamics of super-excited states involving competing ionization, dissociation and relaxation channels is, in any way, interesting by itself.

By combining different spectroscopic techniques (ultrafast pump-probe techniques, frequency sum generation, single molecule spectroscopy, near-field microscopy, photon echo etc.), it is now possible to investigate spectra and dynamics of molecules in well-defined sites at solid surfaces and interfaces, or contained in the supramolecular systems and biological structures.

References

Chapter 1: The Basic Notions

[1]   P. Pringsheim: *Fluorescence and Phosphorescence*, (Interscience 1949)
[2]   G.B. Kistiakovsky, C.S. Parmenter: J. Chem. Phys. **42**, 2942, (1965)
[3]   C.S. Parmenter, H. Poland: J. Chem. Phys. **41**, 655 (1964)
[4]   M. Gouterman: J. Chem. Phys. **36**, 2846 (1962)
[5]   G.W. Robinson, R.P. Frosch: J. Chem. Phys. **37**, 1962 (1962)
[6]   G.R. Hunt, E.F. McCoy, I.G. Ross: Austr. J. Chem. **15**, 591 (1962)
[7]   J.P. Byrne, E.F. McCoy, I.G. Ross: Austr. J. Chem. **18**, 1589 (1965)
[8]   M. Bixon, J. Jortner: J. Chem. Phys. **48**, 715, (1968)
[9]   A. Nitzan, J. Jortner, P.M. Rentzepis: Proc. Roy. Soc. London **A327**, 367 (1972)
[10]   J. Jortner, S. Mukamel: In: *The World of Quantum Chemistry*, ed by R. Daudel, B. Pullman (Reidel Publishing, Dordrecht 1975) p 205
[11]   P. Avouris, W.M. Gelbart, M.A. El-Sayed: Chem. Rev. **77**, 793 (1977)
[12]   S. Mukamel, J. Jortner: In: *Excited States*, vol III, ed by E.C. Lim (Academic Press 1978), p 57
[13]   A. Tramer, R. Voltz: In: *Excited States*, vol IV, ed by E.C. Lim (Academic Press 1979), p 281
[14]   E.S. Medvedev, V.I. Osherev: *Radiationless Transitions in Polyatomic Molecules* (Springer, Berlin Heidelberg New York 1993)
[15]   V. May, O. Kühn: *Charge and Energy Transfer in Molecular Systems* (Wiley-VCH, Berlin 2000)
[16]   W. Heitler: *Quantum Theory of Radiation* (Clarendon, Oxford 1954)
[17]   C. Cohen-Tannoudji, B. Diu, F. Laloë: *Mécanique Quantique* (Hermann, Paris 1973)
[18]   F. Polik, D.R. Guyer, C.B. Moore: J. Chem. Phys. **92**, 3453 (1990)
[19]   W.D. Lawrance, AE.W. Knight: J. Phys. Chem. **89**, 917 (1985)
[20]   A.A. Stuchebrukhov, R.A. Marcus: J. Chem. Phys. (a) **98**, 6044 (1993); (b) **98**, 12491 (1993)
[21]   J. Franck: Trans. Far. Soc. **21**, 536 (1925)

[22]   E.U. Condon:Phys. Rev. **32**, 858 (1928)
[23]   W. Siebrand: J. Chem. Phys. **46**, 440 (1967)
[24]   W. Siebrand: J. Chem. Phys. **47**, 2411 (1967)
[25]   W. Siebrand, D.F. Wlliams: J. Chem. Phys. **49**, 1860 (1968)
[26]   A. Nitzan, S. Mukamel, J. Jortner: J. Chem. Phys. (a) **60**, 3929 (1974); (b) **63**, 200, (1975)
[27]   F. Legay: in *Chemical and Biochemical Applications of Lasers*, vol II, ed by C.B. Moore (Academic Press 1977) p 43
[28]   J.A. Beswick, J. Jortner: J. Chem. Phys. (a) **68**, 2277 (1978); (b) **69**, 512 (1978)
[29]   G.E. Ewing: (a) Chem. Phys. **29**, 253 (1978); (b) J. Chem. Phys. **71**, 3143 (1979); (c) J. Chem. Phys. **72**, 2096 (1979)
[30]   C.S. Parmenter, K.Y. Tang: Chem. Phys. **27**, 127, (1978)
[31]   G. Herzberg, C. Jungen: J. Mol. Spectry **41**, 425, (1972)
[32]   H.J. Korsch, H. Laurent: J. Phys. B, At. Mol. Phys. **14**, 4213, (1981)
[33]   U. Fano: Phys. Rev. **A32**, 617, (1985)

Chapter 2: Experimental Techniques

[1]   G. Scoles: *Atomic and Molecular Beam Methods* (Oxford University Press 1980)
[2]   J.R. Cable, M.J. Tubergen, D.H. Levy: J. Am. Chem. Soc. **111**, 9032 (1989)
[3]   C.S. Parmenter, A.M. White: J. Chem. Phys. **50**, 1631 (1969)
[4]   J.J. Scherer, J.B. Paul, A. O'Keefe, R.J. Saykally: Chem. Rev. **97**, 25 (1997)
[5]   M. Gruebele, G. Roberts, M. Dantus, R.M. Bowman, A.H. Zewail: Chem. Phys. Lett. **166**, 459 (1990)
[6]   V.S. Letokhov: *Laser Photoinization Spectroscopy* (Academic Press, New York 1987)
[7]   R.H. Page, Y.S. Chen, Y.T. Lee: J. Chem. Phys. **88**, 4621 (1988)
[8]   S. Tanabe, T. Ebata, M. Fuji, B. Mikami: Chem. Phys. Lett. **215**, 347 (1997)
[9]   P.J. de Lange, B.J. van der Meer, K E. Drabe, J. Kommandeur, W.L. Meerts, W.A. Majewski: J. Chem. Phys. **86**, 4004 (1987)
[10]   B. Cagnac, S. Haroche, S. Liberman: In: *Frontiers in Laser Spectroscopy* ed by S. Haroche. (North Holland, Amsterdam 1972) p 299
[11]   K.H. Fung, D.A. Ramsay: Mol. Phys. **88**, 9975, (1996)
[12]   G.J. Small (Chap 9), R.I. Personov (Chap 10): In: *Spectroscopy and Excitation Dynamics of Condensed Molecular Systems*, ed by V.M. Agranovich, R.M. Hochstrasser (North-Holland, Amsterdam 1983)
[13]   R.I. Personov, E.I. Al'shits, E.I. Bykovskaja: Opt. Comm. **6**, 169 (1972)
[14]   E. Riedle, H.J. Neusser, E.W. Schlag: Farad. Disc. **75**, 207 (1983)
[15]   M. Lombardi, R. Jost, C. Michel, A. Tramer: Chem. Phys. **57**, 341, 355 (1981)
[16]   H.S. Yoo, M.J. DeWitt, B.H. Pate: J. Phys. Chem. **A 108**, 1348 (2004)
[17]   J.C. Mialocq, T. Gustavsson: In: *New Trends in Fluorescence Spectroscopy*, ed by B. Valeur and J. C. Brochon (Springer, Berlin Heidelberg New York 2001)

[18] R.D. Levine, R.B. Bernstein: *Molecular Reaction Dynamics and Chemical Reactivity* (Oxford University Press 1987)

[19] V.E. Bondybey: In: *Chemistry and Physics of Matrix-Isolated Species* ed by L. Andrews, M. Moskovitz. (Elsevier Berlin 1989)

[20] C. Callegari, K.K. Lehmann, R. Schmied, G. Scoles: J. Chem. Phys. **115**, 10090 (2001)

[21] F. Stiekemeier, A.F. Vilesov: J. Chem. Phys. **115**, 10119, (2001)

[22] R. von Benten, O. Link, B. Abel, D. Schwartzer: J. Phys. Chem. **A 108**, 363 (2004)

[23] W.E. Moerner, M. Orrit: Science **283**, 1670 (1999)

[24] S. Nie, D.T. Chiu, R.N. Zare: Science **266**, 1018 (1986)

[25] J.Y. Huang, Y.R. Shen: In: *Laser Spectroscopy and Photochemistry on Metal Surfaces*, ed by H.L. Dai, W. Ho. (World Scientific, Singapore 1995) p 5

[26] N. Mikami, A. Hiraya, I. Fujiwara, M. Ito: Chem. Phys. Lett. **74**, 531 (1980)

Chapter 3: Intramolecular Vibrational Redistribution

[1]   C.S. Parmenter, A.M. White: J. Chem. Phys. **50**, 1631, (1969)

[2]   A. Frad, F. Lahmani, A. Tramer, C. Tric: J. Chem. Phys. **60**, 4419 (1974)

[3]   M. Stockburger, H. Gattermann, W. Klusmann: J. Chem. Phys. **63**, 4519, 4529 (1975)

[4]   S.M. Beck, D.E. Powers, J.B. Hopkins, R E. Smalley: J. Chem. Phys. (a) **73**, 2019 (1980); (b) **74**, 43 (1981)

[5]   J.B. Hopkins, D.E. Powers, R.E. Smalley: J. Chem. Phys. **72**, 5039 (1980) and the references therein

[6]   D.E. Powers, J.B. Hopkins, R.E. Smalley: J. Chem. Phys. **72**, 5721 (1980)

[7]   J.B. Hopkins, D.E. Powers, R.E. Smalley: J. Chem. Phys. **73**, 4586 (1980)

[8]   P.M. Felker, A.H. Zewail: Chem. Phys. Lett. **102**, 113 (1983)

[9]   W.R. Lambert, P.M. Felker, A.H. Zewail: J. Chem. Phys. **81**, 2217 (1984)

[10] P.M. Felker, A.H. Zewail: J. Chem. Phys. **82**, 2975, (1985)

[11] J.M. Smith, X. Zhang, J.L. Knee: J. Phys. Chem. **99**, 1768 (1995)

[12] M.J. Côté, J.F. Kauffman, P.G. Smith, J.D. McDonald: J. Chem. Phys. **90**, 2865 (1989)

[13] P.G. Smith, J.D. McDonald: J. Chem. Phys. **96**, 7344 (1992)

[14] A. Amirav, U. Even, J. Jortner: Chem. Phys. Lett. **71**, 12 (1980)

[15] A. Amirav, J. Jortner, S. Okajima, E.C. Lim: Chem. Phys. Lett. **126**, 487 (1986)

[16] W.D. Lawrance, A.E.W. Knight: J. Phys. Chem. **89**, 917 (1985)

[17] C.S. Parmenter: Disc. Far. Soc. **75**, 9 (1983)

[18] D.A. Dolson, K.W. Hotzclaw, S H. Lee, S. Munchak, C.S. Parmenter: B. M. Stone, Laser Chem. **2**, 271 (1983)

[19] S. Baskin, T.S. Rose, A.H. Zewail: J. Chem. Phys. **88**, 1458 (1988)

[20] M.J. Côté, J.F. Kauffman, P.G. Smith,, J.D. McDonald: J. Chem. Phys. **90**, 2865 (1989)

[21] M. Castella, P. Millié, F. Piuzzi, J. Caillet, J. Langlet, P. Claverie, A. Tramer: J. Phys. Chem. **93**, 3941, 3949, (1989)

[22] R.J. Babbitt, M.R. Topp: Chem. Phys. Lett. **127**, 111 (1986)

[23] A.L. Motyka, S.A. Wittmeyer, R.J. Babbitt, M.R. Topp: J. Chem. Phys. **89**, 4586 (1988)

[24] D.B. Moss, C.S. Parmenter, G.E. Ewing: J. Chem. Phys. 86, 51 (1987)

[25] P.J. Timbers, C.S. Parmenter, D.B. Moss: J. Chem. Phys., **100**, 1028 (1994)

[26] P.G. Smith,, J.D. McDonald: J. Chem. Phys. **92**, 1004 (1990)

[27] B. Fourmann, C. Jouvet, A. Tramer, J.M. Le Bars, P. Millié: Chem. Phys. **92**, 25 (1985)

[28] J.S. Baskin, M. Dantus, A.H. Zewail: Chem. Phys. Lett. **130**, 473 (1986)

[29] G. Herzberg, *Infrared and Raman Spectra of Polyatomic Molecules* (van Nostrand, N. York 1945)

[30] E. Abramson, R.W. Field, D. Imre, K.K. Innes, J.L. Kinsey: J. Chem. Phys. **83**, 453 (1985)

[31] J.P. Pique, Y.M. Engel, R.D. Levine, Y. Chen, R.W. Field, J.L. Kinsey: J. Chem. Phys. **88**, 5972 (1988)

[32] M.A. Temsamani, M. Herman: J. Chem. Phys. **102**, 6371 (1995)

[33] M.A. Temsamani, M. Herman, S.A.B. Solina, J.P. O'Brien, R.W. Field: J. Chem. Phys. **105**, 11357 (1996)

[34] M.I. El Idrissi, J. Liévin, A. Campargue, M. Herman: J. Chem. Phys. **110**, 2074, (1999)

[35] G. Stewart, J.D. McDonald: J. Chem. Phys. **78**, 3907 (1983)

[36] H.L. Kim, T.J. Kulp, J.D. McDonald: J. Chem. Phys. **87**, 4376 (1987)

[37] H.L. Kim, T.K. Minton, R.S. Ruoff, T.J. Kulp, J.D. McDonald: J. Chem. Phys. **89**, 3955 (1988) and references therein

[38] A. Laubereau, D. von der Linde, W. Kaiser: Phys. Rev. Lett. **28**, 1162 (1972)

[39] D. Bingemann, M.P. Gorman, A.M. King, F.F. Crim: J. Chem. Phys. **107**, 661 (1997)

[40] T. Ebata, A. Iwasaki, N. Mikami, J. Phys. Chem. **A104**, 7974 (2000)

[41] T. Ebata, M. Kayano, S. Sato, N. Mikami: J. Phys. Chem. **A105**, 8623 (2001)

[42] Y. Yamada, T. Ebata, M. Kayano, N. Mikami: J. Chem. Phys. **120**, 7400, (2004)

[43] M. Kayano, T. Ebata, Y. Yamada, N. Mikami: J. Chem. Phys. **120**, 7410, (2004)

[44] H.S. Yoo, M.J. DeWitt, B.H. Pate: (a) J. Phys. Chem. **A108**, 1348 (2004); (b) J. Phys. Chem. **A108**, 1365 (2004)

[45] H.S. Yoo, D.A. McWhorter, B.H. Pate: J. Phys. Chem. **A108**, 1380 (2004)

[46] R. von Benten, O. Link, B. Abel, D. Schwarzer: J. Phys. Chem. **A108**, 325, 363 (2004)

[47] H.J. Bakker: J. Chem. Phys. **94**, 1730 (1991), **98**, 8496 (1993)

[48] J.C. Deak, L.K. Iwaki, D.D. Dlott: (a) J. Phys. Chem. **A102**, 8193 (1998); (b) Chem. Phys. Lett. **293**, 405 (1998); (c) Chem. Phys. Lett. **321**, 419 (2000)

[49] K.V. Reddy, D.F. Heller, M.J. Berry: J. Chem. Phys. **65**, 179 (1976)

[50] R.H. Page, Y.R. Shen, Y.T. Lee: J. Chem. Phys. **88**, 4621 (1988)

[51] D. Bassi, L. Menegotti, S. Oss, M. Scotoni, F. Iachello: Chem. Phys. Lett. **207**, 167 (1993)

[52]   M. Scotoni, C. Leonardi, D. Bassi: J. Chem. Phys. **95**, 8655 (1991)
[53]   A. Callegari, U. Merker, P. Engels, H.K. Srivastava, K.K. Lehmann, G. Scoles: J. Chem. Phys. **113**, 10583 (2000)
[54]   M. Lewerenz, M. Quack: J. Chem. Phys. **88**, 5408, (1988)
[55]   J.E. Baggott, M.C. Chuang, R.N. Zare, HR Dübal, M. Quack: J. Chem. Phys. **82**, 1186 (1985)
[56]   H.R. Dübal, M. Quack: J. Chem. Phys. **81**, 3779, (1984)
[57]   J. Segall, R.N. Zare, H.R Dübal, M. Lewerentz, M. Quack: J. Chem. Phys. **86**, 634 (1987)
[58]   C. Iung, C. Leforestier: J. Chem. Phys. **97**, 2481, (1992)
[59]   O.V. Boyarkin, T.R. Rizzo, D.S. Perry: J. Chem. Phys. **110** (a) 11346 (b) 11359 (1999)
[60]   A. Chirokolava, D.S. Perry, O.V. Boyarkin, M. Schmid, T. R. Rizzo: J. Chem. Phys. **113**, 10068 (2000)
[61]   D. Rueda, O.V. Boyarkin, T.R. Rizzo, I. Mukhopadhyay, D.S. Perry: J. Chem. Phys. **116**, 91 (2002)
[62]   A.M. de Souza, D. Kaur, D.S. Perry: J. Phys. Chem. **88**, 4569 (1988)
[63]   A. McIlroy, D.J. Nesbitt: J. Chem. Phys. **91**, 104, (1989)
[64]   A. McIlroy, D.J. Nesbitt, E.R.Th. Kerstel, B.H. Pate, K.K. Lehmann, G. Scoles: J. Chem. Phys. **100**, 2596 (1994)
[65]   J.E. Gambogi, E.R.Th. Kerstel, K.K. Lehmann, G. Scoles: J. Chem. Phys. **100**, 2612 (1994)
[66]   E.R.Th. Kerstel, K.K. Lehmann, B.H. Pate, G. Scoles: J. Chem. Phys. **100**, 2588 (1994)
[67]   A. Campargue, L. Biennier, A. Garnache, A. Kachanov, D. Romanini, M. Herman: J. Chem. Phys. **111**, 7888, (1999)
[68]   A. McIlroy, D.J. Nesbitt: J. Chem. Phys. **92**, 2229 (1990)
[69]   E.R.Th. Kerstel, K.K. Lehmann, T.F. Mentel, B.H. Pate and G. Scoles: J. Phys. Chem. **95**, 8282 (1991)
[70]   J.E. Gambogi, R.P. L'Esperance, K.K. Lehmann, B.H. Pate, G. Scoles: J. Chem. Phys. **98**, 1116 (1993)
[71]   B.H. Pate, K.K. Lehmann, G. Scoles: J. Chem. Phys. **95**, 3891 (1991)
[72]   T. User, J. T. Hynes, Chem. Phys. **139**, 163, (1989)
[73]   S.M. Lederman, V. Lopez, V. Fairen, G.A. Voth, R.A. Marcus: Chem. Phys. **139**, 171 (1989)
[74]   B. Kuhn, T. R. Rizzo: J. Chem. Phys. **112**, 7461 (2000)
[75]   J. Berkovitz: *Photoabsorption, Photoionization and Photoelectron Spectroscopy* (Academic Press, New York 1979)
[76]   X. Luo, P.R. Fleming, T.R. Rizzo: J. Chem. Phys. **96**, 5659 (1992)
[77]   A. Callegari, J. Rebstein, J.S. Muenter, R. Jost, T.R. Rizzo: J. Chem. Phys. **111**, 123 (1999)
[78]   A. Callegari, J. Rebstein, R. Jost, T.R. Rizzo: J. Chem. Phys. **111**, 7359 (1999)
[79]   M.R. Wedlock, R. Jost, T.R. Rizzo: J. Chem. Phys. **107**, 10344 (1997)
[80]   R.J. Barnes, A. Sinha: J. Chem. Phys. **107**, 3730, (1997)
[81]   H.R. Dübal, F.F. Crim: J. Chem. Phys. **83**, 3863 (1985)

[82]   L. Brouwer, C.J. Cobos, J. Troe, H.R. Dübal, F.F. Crim: J. Chem. Phys. **86**, 6171 (1987)
[83]   I.M. Mills: In: *Molecular Spectroscopy; Modern Research*, ed by N.K. Rao, C. W. Mathews (Academic Press, New York 1972)
[84]   C. Allen, P.C. Cross: *Molecular Vib-rotors* (Wiley, New York 1963)
[85]   G.M. McClelland, G.M. Nathanson, J.H. Frederick, F.W. Farley: In: *Excited States*, vol 7, ed by E.C. Lim, K.K. Innes (Academic Press, New York 1987)
[86]   G.M. Nathanson, G.M. McClelland: J. Chem. Phys. 84, 3170, (1986)
[87]   A.A. Stuchebrukhov, R.A. Marcus: J. Chem. Phys. **98**, 6044 (1993)
[88]   A.A. Stuchebrukhov, A. Mehta, R.A. Marcus: J. Phys. Chem. **98**, 12491 (1993)
[89]   A. Mehta, A.A. Stuchebrukhov, R.A. Marcus: J. Phys. Chem. **99**, 2677 (1995)
[90]   V.S. Letokhov, C.B. Moore: In: C.B. Moore, *Chemical and Biochemical Applications of Lasers*, vol 5, ed by C.B. Moore (Academic Press, New York 1980)
[91]   T. Brixner, N.H. Damrauer, G. Gerber: Adv. Atom. Mol. Opt. Phys. **46**, 1 (2001)

Chapter 4: Vibrational and Rotational Relaxation

[1]   J.O. Hirschfelder, C.F. Curtiss, R.B. Bird: *Molecular Theory of Gases, Liquids* (J. Wiley, New York, 1954)
[2]   D.C. Clary: J. Phys. Chem. (a) **91**, 1718 (1987); (b) **86**, 813 (1987)
[3]   R.D. Levine, R.B. Bernstein; *Molecular Reaction Dynamics, Chemical Reactivity* (Oxford University Press, Oxford, 1987)
[4]   A.J. Caffery: Phys. Chem. Chem. Phys. **6**, 1637, (2004)
[5]   S.P. Phipps, T.C. Smith, G.D. Hager, M.C. Heaven, J.K. McIver, W.G. Rudolph: J. Chem. Phys. **116**, 9281, (2002)
[6]   D.R. Crosley: J. Phys. Chem. **93**, 6273 (1989) and references therein
[7]   S.R. Leone: J. Phys. Chem. Ref. Data 11, 953 (1982) and references therein
[8]   J.P. Toennies: Ann. Rev. Phys. Chem. **27**, 225, (1976)
[9]   D.J. Krajnovich, C.S. Parmenter, D.L. Catlett Jr.: Chem. Rev. **87**, 237 (1987)
[10]   G.W. Flynn, C.S. Parmenter, A.M. Wodtke: J. Phys. Chem. **100**, 12817 (1996)
[11]   D.A.V. Kliner, R.L. Farrow: J. Chem. Phys. **110**, 412 (1999)
[12]   L.S. Dzelzkalns, F. Kaufman: J. Chem. Phys. **79**, 3836 (1983) and references therein
[13]   K.J. Rensberger, J.T. Blair, F. Weinhold, F.F. Crim: J. Chem. Phys. **91**, 1688 (1989)
[14]   H.M. ten Brink, J. Langelaar, R.P.H. Rettschnick: Chem. Phys. Lett. **62**, 263 (1979), **75**, 115 (1981)
[15]   J.D. Tobiason, A.L. Utz, F.F. Crim: J. Chem. Phys. **97**, 7437 (1992), **101**, 1108 (1994)
[16]   M.J. Frost: J. Chem. Phys. **98**, 8572 (1993)
[17]   J. Wu, R. Huang, M. Gong, A. Saury, E. Carrasquillo: J. Chem. Phys. **99**, 6474 (1993)

[18]   F. Temps, S. Halle, P.H. Vaccaro, R.W. Field, J.L. Kinsey: J. Chem. Soc. Faraday Trans. II. **84**, 1457, (1988)
[19]   J.L. Rinnenthal, K.-H. Gericke: J. Chem. Phys. **116**, 9776 (2002)
[20]   A.P. Milce, B.J. Orr: J. Chem. Phys. **104**, 6423, (1996), **106**, 3592 (1997)
[21]   S.M. Clegg, A.B. Burrill, C.S. Parmenter: J. Phys. Chem. **A 102**, 8477 (1998) and references therein
[22]   C.S. Parmenter, K.Y. Tang: Chem. Phys. **27**, 127 (1978)
[23]   D.A. Chernoff, S.A. Rice: J. Chem. Phys. **70**, 2511, 5221 (1979)
[24]   D.L. Catlett Jr., C.S. Parmenter: J. Phys. Chem. **95**, 2864 (1991), **98**, 3263 (1994), **99**, 7371 (1995)
[25]   G. Hall, C.F. Giese, W.R. Gentry: J. Chem. Phys. **83**, 5343 (1985)
[26]   M.C. Wall, A.E. Lernoff, A.S. Mullin: J. Chem. Phys. **111**, 7373 (1999), M.S. Elioff, M. Fraelich, R.L. Sansom, A. S. Mullin, J. Chem. Phys. **111**, 3517, (1999)
[27]   E.T. Sevy, S.M. Rubin, Z. Lin, G.W. Flynn: J. Chem. Phys. **113**, 4912 (2000)
[28]   V.S. Vikhrenko, C. Heidelbach, D. Schwarzer, V.B. Nemtsov, J. Schroeder: J. Chem. Phys. **110**, 5273 (1999)
[29]   G. Herzberg: *Infrared and Roman Spectra of Polyatomic Molecules* (van Nostrand, Princeton, 1945)
[30]   K.M. Beck, M.T. Berry, M.B. Brustein, M.I. Lester: Chem. Phys. Lett. **162**, 203 (1989)
[31]   Z.S. Huang, K.W. Jucks, R.E. Miller: J. Chem. Phys. **85**, 6905 (1986)
[32]   G.T. Fraser, A.S. Pine: J. Chem. Phys. **85**, 2502, (1986)
[33]   W. Sharfin, K.E. Johnson, L. Wharton, D.H. Levy: J. Chem. Phys. **71**, 1292 (1979), J.E. Kenny, K.E. Johnson, W. Sharfin, D.H. Levy: J. Chem. Phys. **72**, 1109, (1980)
[34]   J.A. Beswick, J. Jortner: J. Chem. Phys. 68, 2277 (1978), **69**, 512 (1978)
[35]   A. Burroughs, M.C. Heaven: J. Chem. Phys. **114**, 7027 (2001)
[36]   C.M. Lovejoy, D.D. Nelson, D.J. Nesbitt: J. Chem. Phys. **87**, 5621 (1987), **89**, 7180 (1988)
[37]   D.J. Nesbitt, C.M. Lovejoy: J. Chem. Phys. **99**, 7716 (1990)
[38]   J.H. Shorter, M.P. Carcassa, D.S. King: J. Chem. Phys. **97**, 1824 (1992)
[39]   M.D. Wheeler, M.W. Todd, D.T. Anderson, M. Lester: J. Chem. Phys. **110**, 6732 (1999)
[40]   P.J. Krause, D.C. Clary, D.A. Anderson, M.W. Todd, R.L. Schwartz, M.I. Lester: Chem. Phys. Lett. **294**, 518 (1998)
[41]   M.D. Wheeler, M. Tsiouris, M. Lester, G. Lendvay: J. Chem. Phys. **112**, 6590 (2000)
[42]   M.W. Todd, D.T. Anderson, M. Lester: J. Chem. Phys. **110**, 6532 (1999)
[43]   E.J. Bohac, R.E. Miller: Phys. Rev. Lett. **71**, 54 (1993)
[44]   B.V. Pond, M.I. Lester: J. Chem. Phys. **118**, 2223, (2003)
[45]   D.S. Tinti, G.W. Robinson: J. Chem. Phys. **63**, 2842, (1975)
[46]   H. Dubost, R. Charneau: Chem. Phys. **12**, 407 (1976)
[47]   A. Salloum, H. Dubost: Chem. Phys. **189**, 179 (1994)
[48]   L.E. Brus, V.E. Bondybey: J. Chem. Phys. **63**, 786 (1975)

[49] J.M. Wiesenfeld, C.B. Moore: J. Chem. Phys. **70**, 930 (1979)

[50] F. Legay: In: *Chemical and Biochemical Applications of Lasers*, vol II, ed by C.B. Moore. (Academic Press 1977) p 43

[51] K.F. Freed, H. Metiu: Chem. Phys. Lett. **48**, 262 (1977)

[52] M. Berkowitz, R.B. Gerber: Phys. Rev. Lett. **39**, 1000 (1977)

[53] S.M. Bellm, W.D. Lawrance: J. Chem. Phys. **118**, 2581 (2003)

[54] D.V. Brumbaugh, J.E. Kenny, D.H. Levy: J. Chem. Phys. **78**, 3415 (1983)

[55] M. Heppener, A.G. M. Kunst, D. Bebelaar, R.P H. Rettschnick: J. Chem. Phys. **83**, 5341 (1985)

[56] R.M. Nimlos, M.A. Young, E.R. Bernstein, D.F. Kelley: J. Chem. Phys. **91**, 5268 (1989)

[57] M.F. Hineman, S.K. Kim, E.R. Bernstein, D.F. Kelley: J. Chem. Phys. **96**, 4904 (1992)

[58] D.F. Kelley, E.R. Bernstein: J. Phys. Chem. **90**, 5164 (1986)

[59] P.M. Weber, S.A. Rice: J. Chem. Phys. **88**, 6120 (1988)

[60] K.W. Butz, D.L. Catlett, G.E. Ewing, D. Krajnovich, C.S. Parmenter: J. Phys. Chem. **90**, 3533 (1986)

[61] O. Hye-Keun, C.S. Parmenter, M.-C. Su: Ber. Bunsenges. **92**, 253 (1988)

[62] B.A. Jacobson, S. Humphrey, S.A. Rice: J. Chem. Phys. **89**, 5624 (1988)

[63] G. Lembach, B. Brutschy: J. Phys. Chem. **A 102**, 6068, (1997)

[64] P. Asselin, B. Dupuis, J.P. Perchard, P. Soulard: Chem. Phys. Lett. **268**, 265 (1997)

[65] P. Asselin, P. Soulard, M.E. Alikhani, J.P. Perchard: Chem. Phys. **249**, 73 (1999)

[66] A.M.E. Giebels, M.A.F.H. van der Broek, M.F. Kropman, H.J. Bakker: J. Chem. Phys. **112**, 5127 (2000)

[67] M. Broquier, F. Lahmani, A. Zehnacker-Rentien, V. Brenner, Ph. Millié, A. Peremans: J. Phys. Chem. **A 105**, 6841 (2001)

[68] Y. Yamada, T. Ebata, M. Kayano, N. Mikami: J. Chem. Phys. **120**, 7400 (2004)

[69] M. Kayano, T. Ebata, Y. Yamada, N. Mikami: J. Chem. Phys. **120**, 7410 (2004)

[70] B.I. Stepanov: Nature **157**, 808 (1946)

[71] M.V. Volkenshtein, M.A. Eliashevich, B.I. Stepanov: *Molecular Vibrations*, vol 2 (in Russian) (Moscow 1949) p 250

[72] R. Laenen, K. Simeonidis: Chem. Phys. Lett. **299**, 589 (1999)

[73] R. Laenen, G.M. Gale, N. Lascoux: J. Phys. Chem. **A 103**, 10708 (1999)

[74] A. Staib, J.T. Hynes: Chem. Phys. Lett. **204**, 197, (1993)

[75] H. Graener, T. Lösch, A. Laubereau: J. Chem. Phys. **93**, 5365 (1990)

[76] R. Laenen, C. Rauscher, A. Laubereau: (a) J. Phys. Chem. **A101**, 3201 (1997); (b) Chem. Phys. Lett. **283**, 7 (1998)

[77] K.J. Gaffney, P.H. Davis, I.R. Piletic, N.E. Levinger, M.D. Fayer: J. Phys. Chem. **A 106**, 12012 (2002)

[78] R.N. Schwartz, Z.I. Slavsky, K.F. Herzfeld: J. Chem. Phys. **20**, 1591 (1952)

[79] F.I. Tanczos: J. Chem. Phys. **25** 439 (1956)

[80] A. Miklavc, S. Fischer: J. Chem. Phys. **44**, 209 (1976)

[81] S.G. Fischer, A. Irgens-Defregger: J. Phys. Chem. **87**, 2054 (1983)

[82] D.B. McDonald, S.A. Rice: J. Chem. Phys. **74**, 4918 (1981)

[83]   S. L. Thompson, J. Chem. Phys. **49**, 3400 (1968)
[84]   H. Abdel-Halim, G. Ewing: J. Chem. Phys. **82**, 5442 (1985)
[85]   A.D. Abbate, C.B. Moore: J. Chem. Phys. **82**, 1255, 1263 (1985), **83**, 975 (1985)
[86]   H. Chabbi, B. Gauthier-Roy, A.M. Vasserot, L. Abouaf-Marguin: J. Chem. Phys. **117**, 4436 (2002)
[87]   C. Jouvet, M. Sulkes, S.A. Rice: J. Chem. Phys. **78**, 3935 (1983)
[88]   N. Halberstadt, B. Soep: Chem. Phys. Lett. **87**, 109 (1982)
[89]   N. Halberstadt, B. Soep: J. Chem. Phys. **80**, 2340 (1984)
[90]   L. Lapoerre, D. Frye, H.-L. Dai: J. Chem. Phys. **96**, 2703 (1992)
[91]   R.L. Rosman, S.A. Rice: J. Chem. Phys. **86**, 3292 (1987)
[92]   E.R. Waclawik, W.D. Lawrance: J. Chem. Phys. **102**, 2780 (1995), **109**, 5921 (1998)
[93]   Mudjijono, W.D. Lawrance: J. Chem. Phys. **109**, 6736 (1998) and references therein
[94]   S.H. Kable, A.E.W. Knight: J. Chem. Phys. **93**, 3151 (1990)

Chapter 5: Electronic Relaxation

[1]   W. Siebrand: J. Chem. Phys. **46**, 440 (1967); **47**, 2411 (1967)
[2]   W. Siebrand, D.F. Wlliams: J. Chem. Phys. **49**, 1860 (1968)
[3]   S.P. McGlynn, T. Azumi, M. Kinoshita: *Molecular Spectroscopy of the Triplet State*, (Prentice Hall, Englewood Cliffs, New Jersey 1969)
[4]   M. Kasha: Disc. Far. Soc. **9**, 14 (1950)
[5]   M. Beer, H.C. Longuet-Higgins: J. Chem. Phys. **23**, 1390 (1955)
[6]   T.M. Woudenberg, S.K. Kulkarni,, J.E. Kenny: J. Chem. Phys. **89**, 2789 (1988)
[7]   D.R. Demmer, J.W. Hager, G.W. Leach, S.C. Wallace: Chem. Phys. Lett. **136**, 329 (1987)
[8]   S.Z. Levine, A.R. Knight, R.P. Steer: Chem. Phys. Lett. **29**, 73 (1974)
[9]   A.M. Warsylewicz, K.J. Falk, R.P. Steer: Chem. Phys. Lett. **352**, 48 (2002)
[10]   M. Kawasaki, K. Kasatani, H. Sato: Chem. Phys. **94**, 179 (1985)
[11]   M. Mahaney, J.R. Huber: Chem. Phys. **9**, 371 (1975)
[12]   A. Wittmeyer, A.J. Kaziska, M.I. Shchuka, A.L. Motyka, M.R. Topp: Chem. Phys. Lett. **151**, 384 (1988)
[13]   C.J. Ho, A.L. Motyka, M.R. Topp: Chem. Phys. Lett. **158**, 51 (1989)
[14]   J.P. Maier, O. Marthaler: Chem. Phys. **32**, 419 (1978)
[15]   J.P. Maier, F. Thommen: Chem. Phys. **57**, 319 (1981)
[16]   T. Pino, N. Boudin, P. Bréchignac: J. Chem. Phys. **111**, 7337 (1999)
[17]   A. Amirav, C. Horwitz, J. Jortner: J. Chem. Phys. **88**, 3092 (1988)
[18]   A. Amirav, J. Jortner: Chem. Phys. Lett. **132**, 335 (1986)
[19]   A. Kearbell, F. Wilkinson: J. Chim. Phys. **20**, 125 (1970)
[20]   T. Deinum, C.J. Werkhoven, J. Langelaar, R.P.H. Rettschnick, J.D.W. Van Voorst: Chem. Phys. Lett. **27**, 552 (1974)

[21] H. Baba, A. Nakajima, M. Aoi, K. Chihara: (a) J. Chem. Phys. **55**, 2433 (1971); (b) Mol. Phys. **88**, 9975 (1996)

[22] S.M. Beck, J.B. Hopkins, D.E. Powers, R E. Smalley: J. Chem. Phys. **74**, 43 (1981)

[23] C.A. Langhoff, G.W. Robinson: Chem. Phys. **6**, 34 (1974)

[24] S.M. Beck, D.E. Powers, J.B. Hopkins, R.E. Smalley: J. Chem. Phys. **73**, 2019 (1980)

[25] A. Jablonski: Nature **131**, 839 (1933), Z. Phys. **94**, 38 (1935)

[26] R. van der Werf, D. Zevenhuijzen, J. Jortner: Chem. Phys. **27**, 319 (1978)

[27] R.W. Field, B.G. Wicke, J.D. Simmons, S.G. Tilford: J. Mol. Spectr. **44**, 383 (1972) and references therein

[28] M. Lavollée, A. Tramer: Chem. Phys. Lett. **47**, 523 (1977)

[29] M. Lavollée, A. Tramer: Chem. Phys. **45**, 45 (1976)

[30] B. Girard, N. Billy, J. Vigué, J.C. Lehmann: Chem. Phys. Lett. **92**, 615 (1982)

[31] M. Kawasaki, K. Kasatani, H. Sato: Chem. Phys. **94**, 179 (1985)

[32] D.J. Clouthier, G. Huang, A.J. Merer: J. Chem. Phys. **97**, 1630 (1992)

[33] J.R. Dunlop and D.J. Clouthier: J. Chem. Phys **93**, 6371, (1990)

[34] K.H. Fung, D.A. Ramsay: (a) J. Phys. Chem. **88**, 395 (1984); (b) Mol. Phys. **88**, 9975, (1996)

[35] I. Burak, J.W. Hepburn, N. Sivakumar, G.E. Hall, G. Chawla, P.L. Houston: J. Chem. Phys. **86**, 1258 (1987)

[36] C. Michel, A. Tramer: Chem. Phys. **42**, 315 (1979)

[37] F.W. Birss, J.M. Brown, A.R.H. Cole, A. Lofthus, S.L.N. G. Krishnamachari, G.A. Osborne, J. Paldus, D.A. Ramsay, L. Watman: Can. J. Phys. **48**, 1230 (1970)

[38] C. Michel, M. Lombardi, R. Jost: Chem. Phys. **109**, 357 (1986)

[39] M. Lombardi: In: *Excited States*, vol 7, ed by E.C. Lim. (Academic Press, New York 1988) p 163

[40] J. Heldt, Ch. Ottinger, A.F. Vilesov, T. Winkler: J. Phys. Chem. **A101**, 740 (1997)

[41] J. Chaiken, M. Gurnick, J.D. McDonald: J. Chem. Phys. **74**, 116 (1981)

[42] J. Chaiken, J.D. McDonald: J. Chem. Phys. **77**, 669 (1982)

[43] K.F. Greenough, A.B.F. Duncan: J. Am. Chem. Soc. **83**, 555 (1961)

[44] R. Kullmer, W. Demtröder: J. Chem. Phys. **84**, 3672 (1986)

[45] S.C. Bae, H.S. Yoo, J.K. Koo: J. Chem. Phys. **109**, 1251 (1998)

[46] A.E. Douglas: J. Chem. Phys. **45**, 1007 (1966)

[47] F. Lahmani, A. Tramer, C. Tric: J. Chem. Phys. **60**, 4419, 4431 (1974)

[48] K.G. Spears, M. El-Manguch: Chem. Phys. **24**, 65 (1977)

[49] R. van der Werff, J. Kommandeur: Chem. Phys. **16**, 125 (1976)

[50] J.A. Honings, W.A. Majewski, Y. Matsumoto, D.W. Pratt, W.L. Meerts: J. Chem. Phys. **89**, 1813 (1988)

[51] N. Ohta, M. Fujita, H. Baba: Chem. Phys. Lett. **135**, 330 (1987)

[52] W.M. van Herpen, P.A.M. Uijt de Haag, W.L. Meerts: J. Chem. Phys. **89**, 3939 (1988)

[53] W.D. Lawrance, A.E.W. Knight: J. Phys. Chem. **89**, 917 (1985)

[54] W.M. van Herpen, W. L. Meerts, K.E. Drabe, J. Kommandeur: J. Chem. Phys. **86**, 4396 (1987)

[55] N. Ohta: J. Phys. Chem. **100**, 7298 (1996) and references therein

[56] R. Carter, H. Bitto, J.R. Huber: J. Chem. Phys. **102**, 5890 (1995)

[57] G.J. van der Meer, H.Th. Jonkman, G.M. ter Horst, J. Kommandeur: J. Chem. Phys. **76**, 2099 (1982)

[58] P.M. Felker, A.H. Zewail: Chem. Phys. Lett. **128**, 221 (1986)

[59] A. Amirav, J. Jortner: J. Chem. Phys. **84**, 1500 (1986)

[60] N. Ohta, O. Sekiguchi, H. Baba: J. Chem. Phys. **88**, 68 (1988)

[61] N. Ohta, T. Takemura, H. Baba: J. Phys. Chem. **92**, 5554 (1988)

[62] Y. Matsumoto, L.H. Spangler, D.W. Pratt: J. Chem. Phys. **80**, 5539 (1984)

[63] Y. Matsumoto, D.W. Pratt: J. Chem. Phys. **81**, 573 (1984)

[64] N. Ohta, T. Takemura: Chem. Phys. Lett **169**, 611 (1990)

[65] A. Ali, G. Jihua, P.J. Dagdigian: J. Chem. Phys. **87**, 2045 (1987) and references therein

[66] D.H. Katayama, T.A. Miller, V.E. Bondybey: J. Chem. Phys. **72**, 5439 (1980)

[67] D.H. Katayama: Phys. Rev. Lett. **54**, 657 (1985)

[68] P.R. Harrowell, K.F. Freed: J. Chem. Phys. **83**, 6288 (1985)

[69] H.D. Mettee: J. Chem. Phys. **49**, 1784 (1968)

[70] A.E.W. Knight, J.T. Jones, C.S. Parmenter: J. Phys. Chem. **87**, 973 (1983)

[71] D. Grimbert, M. Lavollée, A. Nitzan, A. Tramer: Chem. Phys. Lett, **57**, 45 (1978)

[72] C. Jouvet, B. Soep: J. Chem. Phys. **73**, 4127 (1980)

[73] M.S. Mangir, H. Reisler, C. Wittig: J. Chem. Phys. **73**, 829, 2280 (1980)

[74] F. Lahmani: J. Phys. Chem. **80**, 2623 (1976)

[75] W.G. Lawrence, Y. Chen, M.C. Heaven: J. Chem. Phys. **107**, 7163 (1997)

[76] Y. Chen, M.C. Heaven: J. Chem. Phys. **112**, 7416 (2000)

[77] V.E. Bondybey, A. Nitzan: Phys. Rev. Lett. **38**, 889 (1977)

[78] G. Herzberg: *Molecular Spectra and Molecular Structure* vol 1 (van Nostrand, Princeton, New Jersey 1950)

[79] R.S. Mulliken: J. Chem. Phys. **33**, 247 (1960)

[80] L.D. Landau: Phys. Z. Sowjetunion **2**, 46 (1932)

[81] C. Zener: Proc. Roy. Soc. (London) **A137**, 696 (1932)

[82] E.J. Heller, R.C. Brown: J. Chem. Phys. **79**, 3336 (1983)

[83] E.E. Nikitin, S.A. Umianski: *Theory of Slow Atomic Collisions* (Springer, Berlin Heidelberg New York 1984)

[84] C. Zhu, H. Nakamura: J. Chem. Phys. **106**, 2599 (1997) and references therein

[85] T.J. Park, J.C. Light: J. Chem. Phys. **85**, 5870 (1986)

[86] U. Manthe, H. Köppel: J. Chem. Phys. **93**, 345 (1990)

[87] M. Broyer, J. Vigué, J.C. Lehmann: J. Chem. Phys. **64**, 4793 (1976)

[88] K. Sakurai, G. Capelle, H.P. Broida: J. Chem. Phys. **54**, 1220 (1971)

[89] E. Martinez, M.T. Martinez, M.P. Puyelo, F. Castano: J. Mol. Structure **175**, 7 (1988)

[90] H. Lefebvre-Brion, R.W. Field: *Spectroscopy and Dynamics of Diatomic Molecules* (Academic Press, New York 2004)

[91]   P. Ho, D. J. Bamford, F.J. Buss, Y.T. Lee, C.B. Moore: J. Chem. Phys. 76, 3630 (1982)

[92]   J.C. Weisshaar, C.B. Moore: J. Chem. Phys. 70, 5135 (1979), 75, 5415 (1980)

[93]   W.F. Polik, D.R. Guyer, C.B. Moore: J. Chem. Phys. 92, 3453 (1990) and references therein

[94]   M. Gruebele, G. Roberts, M. Dantus, R.M. Bowman, A.H. Zewail: Chem. Phys. Lett. 166, 459 (1990)

[95]   L. Krim, B. Soep, J. P. Visticot: J. Chem. Phys. 103, 9589 (1995)

[96]   M. Dantus, M.J. Rosker, A.H. Zewail: J. Chem. Phys. 89, 6127 (1989)

[97]   T.S. Rose, M.J. Rosker, A.H. Zewail: J. Chem. Phys. 91, 7415 (1989)

[98]   S. H. Schaeffer, D. Bonder, E. Tiemannn, Chem. Phys. 89, 65 (1984)

[99]   D.R. Yarkony: Rev. Mod. Phys. 68, 985 (1996) and references therein

[100]   W. Domcke, D.R. Yarkony, H. Köppel: *Conical Intersection* (World Scientific, Singapore 2004)

[101]   C.S. Parmenter, A.M. White: J. Chem. Phys. 50, 1631 (1969)

[102]   T.A. Stephenson, S.A. Rice: J. Chem. Phys. 81, 1073 (1984)

[103]   G.L. Loper, E.K.C. Lee: Chem. Phys. Lett. 13, 140 (1972)

[104]   M. Jacon, C. Lardeux, R. Lopez-Delgado, A. Tramer: Chem. Phys. Lett. 24, 145 (1977)

[105]   C.-S. Huang, J.C. Hsieh, E.C. Lim: Chem. Phys. Lett. 37, 349 (1976)

[106]   C.E. Otis, J.L. Knee, P.M. Johnson: J. Chem. Phys. 78, 2091 (1983); (c) J. Phys. Chem. 87, 2232 (1983)

[107]   T.G. Dietz, M.A. Duncan, R.E. Smalley: J. Chem. Phys. 76, 1227 (1982)

[108]   M.A. Duncan, T.G. Dietz, M.G. Liverman , R.E. Smalley: J. Phys. Chem. 85, 7 (1981)

[109]   T.G. Dietz, M.A. Duncan, A.C. Pulu, R.E. Smalley: J. Phys. Chem. 86, 4026 (1982)

[110]   O. Sneh, O. Cheshnovsky: J. Chem. Phys. 96, 8095 (1992)

[111]   I. Becker, O. Cheshnovsky: J. Chem. Phys. 101, 3649 (1994)

[112]   E. Riedle, H.J. Neusser, E.W. Schlag: Phil. Trans. Roy. Soc. A 332, 189 (1990)

[113]   U. Schubert, E. Riedle, H.J. Neusser: J. Chem. Phys. 84, 5326, 6182 (1986)

[114]   E. Riedle, H.J. Neusser: J. Chem. Phys. 80, 4686 (1984)

[115]   U. Schubert, E. Riedle, H.J. Neusser, E.W. Schlag,: J. Chem. Phys. 84, 6182 (1986)

[116]   E. Riedle, Th. Weber, U. Schubert, H.J. Neusser, E.W. Schlag: J. Chem. Phys. 93, 967 (1990)

[117]   R.S. Friedman, I. Podzielinski, L.S. Cederbaum, V.M. Ryaboy, N. Moiseyev: J. Phys. Chem. A 106, 4320 (2002)

[118]   R.P. Krawczyk, K. Maalsch, G. Hohlneicher, R.C. Gillen, W. Domcke: Chem. Phys. Lett. 320, 535 (2000)

[119]   E.W.G. Diau, S. De Feyter, A.H. Zewail: J. Chem. Phys. 110, 9785 (1999)

[120]   M.J. Bearpark, F. Bernardi, D. Clifford, M. Olivucci, M.A. Robb, B.R. Smith, T. Vren: J. Am. Chem. Soc. 118, 169 (1996)

[121]   A. Ferretti, A. Lami, G. Villani: Chem. Phys. 196, 447, (1995)

[122] A. Raab, G.A. Worth, H.D. Meyer, L.S. Cederbaum: J. Chem. Phys. **110**, 936 (1999)

[123] V. Stert, P. Farmanara, W. Radloff: J. Chem. Phys. **112**, 4460 (2000)

[124] W. Fuss, W.E. Schmidt, S.A. Trushin: Chem. Phys. Lett. **342**, 91 (2001)

[125] W. Fuss, T. Schikarski, W.E. Schmid, S.A. Trushin, P. Herring, K.L. Kompa: J. Chem. Phys. **106**, 2205 (1997)

[126] K. Ohta, Y. Naitoh, K. Saitow, K. Tominaga, N. Hiropa, K. Yoshihara: Chem. Phys. Lett. 256, 629 (1996)

[127] B. Ostojic, W. Domcke: Chem. Phys. **269**, 1 (2001)

[128] M. Garavelli, P. Celani, F. Bernardi, M.A. Robb, M. Olivucci: J. Am. Chem. Soc. **119**, 11487 (1997)

[129] G. Stock, W. Domcke: J. Opt. Soc. Am. **B7**, 1970 (1990)

[130] G. Orlandi, F. Zerbetto, M.Z. Zgierski: Chem. Rev. **91**, 867 (1991) and references therein

[131] F. Zerbetto, M.Z. Zgierski: J. Chem. Phys. **93**, 1235 (1990)

[132] B.E. Kohler: Chem. Rev. **93**, 41 (1993)

[133] P.G. Wilkinson, R.S. Mulliken: J. Chem. Phys. **23**, 1895 (1955)

Chapter 6: The Electron and Proton Transfer

[1]   M.J. Weaver: Chem. Rev. **82**, 463 (1992)

[2]   E.W. Schlag, R. Levine: J. Phys. Chem. **96**, 10608 (1992)

[3]   M.D. Newton: Chem. Rev. **91**, 767 (1991)

[4]   P.F. Barbara, T.J. Meyer, M.A. Rattner: J. Phys. Chem. **100**, 13148 (1996)

[5]   M. Bixon, J. Jortner: Adv. Chem. Phys. **106**, 1 (1999)

[6]   S. Scheiner: J. Phys. Chem. **A 104**, 5898 (2000)

[7]   A. Douhal, F. Lahmani, A.H. Zewail: Chem. Phys. **207**, 477 (1996)

[8]   S.J. Formosinho, L. Arnaut: J. Photochem. Photobiol. **A75**, 1, 21 (1993)

[9]   R.A. Marcus, N. Sutin: Biochim. Biophys. Acta **811**, 265 (1985)

[10] M.N. Paddon-Row: Acc. Chem. Res. **27**, 18 (1994)

[11] b) K.D. Jordan, M.N. Paddon-Row: Chem. Rev. **92**, 400 (1992)

[12] R.S. Mulliken, W.B. Person: *Molecular Complexes* (Wiley-Interscience, New York 1969)

[13] B. Wegewijs, J.W. Verhoeven: Adv. Chem. Phys **106**, 221 (1999)

[14] J.M. Warman, M.P. de Haas, J.W. Verhoeven, M.N. Paddon-Row: Adv. Chem. Phys **106**, 571 (1999)

[15] J.A. Syage, P.M. Felker, A.H. Zewail: J. Chem. Phys. **81**, 2233 (1984)

[16] W. Liptay: In: *Excited States*, vol 1, ed by E.C. Lim (Academic Press, N. York 1974), p 129

[17] K. Rotkiewicz, K.H. Grellmann, Z.R. Grabowski: Chem. Phys. Lett. (a) **19**, 315 (1973); (b) **21**, 212 (1973)

[18] S. Zilberg, Y. Haas: J. Phys. Chem. A**106**, 1 (2002)

[19] D. Pines, E. Pines, W. Rettig: J. Phys. Chem. **A107**, 236 (2003)

198   References

[20]   R.A. Rijkenberg, D. Bebelaar, W.J. Buma, J.W. Hofstraat: J. Phys. Chem. **A 106**, 2446 (2002)
[21]   Z.R. Grabowski, K. Rotkiewicz, W. Rettig: Chem. Rev **103**, 3899 (2003)
[22]   J. Prochorow, J. Lumin:. **9**, 131 (1974)
[23]   T.D. Russell, D.H. Levy: J. Phys. Chem. **86**, 2718 (1982)
[24]   J. Prochorow, A. Tramer: J. Chem. Phys. **44**, 4545 (1966)
[25]   M. Kroll: J. Am. Chem. Soc. **90**, 1097 (1968)
[26]   I.V. Rubtsov, K. Yoshihara: J. Phys. Chem. **A 103**, 10202 (1999)
[27]   K. Wynne, C. Galli, R.M. Hochstrasser: J. Chem. Phys. **100**, 4797 (1994)
[28]   K. Wynne, G.D. Reed, R.M. Hochstrasser: J. Chem. Phys. **105**, 2287 (1996)
[29]   M.H. Hui, W.R. Ware: J. Am. Chem. Soc. **98**, 4718 (1977)
[30]   A. Amirav, M. Castella, F. Piuzzi, A. Tramer: J. Phys. Chem. **92**, 5500 (1988)
[31]   T. Asahi, N. Mataga: J. Phys. Chem. **93**, 6575 (1989)
[32]   S. Ojima, H. Miyasaka, N. Mataga, J. Phys: Chem. **94**, 4147, 7534 (1990)
[33]   T. Asahi, N. Mataga: J. Phys. Chem. **95**, 1956 (1991)
[34]   P.Y. Cheng, D. Zhong, A.H. Zewail: J. Chem. Phys. **103**, 5153 (1995)
[35]   P.Y. Cheng, D. Zhong, A.H. Zewail: J. Chem. Phys. **105**, 6216 (1996)
[36]   G. DeBoer, J.W. Burnett, A. Fujimoto, M A. Young: J. Phys. Chem. **100**, 14882 (1996)
[37]   D. Zhong, T.M. Bernhardt, A.H. Zewail: J. Phys. Chem. **A103**, 10093 (1999)
[38]   H. Knibbe, D. Rehm, A. Weller: Ber. Bunsenges. **72**, 257 (1968)
[39]   H. Stärk, R. Mitzkus, H. Meyer, A. Weller: Appl. Phys. **B30**, 153 (1983)
[40]   A. Tramer, V. Brenner, P. Millié, F. Piuzzi: J. Phys. Chem. **A 102**, 2798 (1998)
[41]   V. Brenner, P. Millié, F. Piuzzi, A. Tramer: J. Chem. Soc. Far. Trans. **93**, 3277 (1997) and references therein
[42]   H. Saigusa, M. Itoh, M. Baba, I. Hanazaki: J. Chem. Phys. **86**, 2588 (1987)
[43]   Q.-H. Xu, G.D. Scholes, M. Yang, G.R. Fleming: J. Phys. Chem. **A 103**, 10348 (1999)
[44]   Y. Nagasawa, A.P. Yartsev, K. Tominaga, P.B. Bisht, A.E. Johnson, K Yoshihara: J. Phys. Chem. **99**, 653 (1996)
[45]   I.V. Rubtsov, H. Shirota: K. Yoshihara, J. Phys. Chem. **A103**, 1801 (1999)
[46]   M. Seel, S. Engleitner, W. Zinth: Chem. Phys. Lett. **275**, 363 (1997)
[47]   S. Engleitner, M. Seel, W. Zinth: J. Phys. Chem. **A 103**, 3013 (1999)
[48]   D.H. Levy: Adv. Chem. Phys. **106**, 203 (1999)
[49]   B. Wegewijs, thesis; Amsterdam (1994)
[50]   M. Polanyi, *Atomic Reactions* (Williams, London 1932)
[51]   XY. Lauteslager, I.H.M. van Stokkum, H.J. van Ramesdonk, A.M. Brouwer, JW. Verhoeven: J. Phys. Chem. **A 103**, 653 (1998)
[52]   H.M. McConnell: J. Chem. Phys. **35**, 508 (1961)
[53]   J.M. Warman, K.J. Smit, S.A. Jonker, J.W. Verhoeven, H. Oevering, J. Kroon, M.N. Paddon-Row, A.M. Oliver: Chem. Phys. **170**, 359 (1993)
[54]   A.Weller, Z. Elektrochem : **60**, 1144 (1956)
[55]   A. Weller: Disc. Far. Soc. **39**, 183 (1965)
[56]   J. Goodman, L. E. Brus: J. Am. Chem. Soc. **100**, 7472 (1978)
[57]   T. Elsaesser, W. Kaiser: Chem. Phys. Lett. **128**, 231 (1986)

[58] L. Helmbrook, J.E. Kenny, B.E. Kohler, G.W. Scott: J. Phys. Chem. **87**, 280 (1983)

[59] J.L. Herek, S. Pedersen, L. Banares, A.H. Zewail: J. Chem. Phys. **97**, 9046 (1992)

[60] N. Shida, J. Almlöf, P.F. Barbara: J. Phys. Chem. **95**, 10457 (1991)

[61] M. Rini, A. Kummrow, J. Dreyer, P. Hamm, E.T.J. Nibbering, T. Elssaesser: Chem. Phys. Lett. **354**, 256 (2002)

[62] A. Douhal, F. Lahmani, A. Zehnacker-Rentien, F. Amat-Guerri: J. Phys. Chem. **A 98**, 12198 (1994)

[63] C. Lu, R.R. Hsieh, I. Lee, P. Cheng: Chem. Phys. Lett. **310**, 103 (1999)

[64] C. Chudoba, E. Riedle, M. Pfeiffer, T. Elsaesser: Chem. Phys. Lett. **263**, 622 (1996)

[65] S. Lochbrunner, A.J. Wurzer, E. Riedle: J. Chem. Phys. **112**, 10699 (2000)

[66] C. Su, J.Y. Lin, R.R. Hsieh, P.Y. Cheng: J. Phys. Chem **A 106**, 11997 (2002)

[67] A. Douhal: Science **276**, 221 (1997)

[68] W. Frey, F. Laermer, T. Elsaesser: J. Phys. Chem. **95**, 10391 (1991)

[69] A. Douhal, F. Lahmani, A. Zehnacker-Rentien: Chem. Phys. **178**, 493 (1993)

[70] A.L. Sobolewski, W. Domcke: Chem. Phys. Lett. **300**, 533 (1999)

[71] B.J. Schwartz, L.A. Peteanu, C.B. Harris: J. Phys. Chem. **96**, 3591 (1992)

[72] A. Mühlfordt, T. Bultmann, N.P. Ernsting, B. Dick: Chem. Phys. **181**, 447 (1994)

[73] A. Ito, Y. Fujiwara, M. Itoh: J. Chem. Phys. **96**, 7474, (1992)

[74] M. Wiechmann, H. Port, F.L Laermer, W. Frey, T. Elsaesser: Chem. Phys. Lett. **165**, 28 (1990)

[75] T. Förster: Z. Elektrochem. **54**, 42 (1950)

[76] A. Weller: *Progress in Reaction Kinetics*, vol 1, (1961) p 176

[77] G. Granucci, J.T. Hynes, P. Millié, T-H. Tran-Thi: J. Amer. Chem. Soc. **122**, 12243 (2000)

[78] N. Mataga, T. Kubota, *Molecular Interactions and Electronic Spectra*, (Fekker, New York 1970) and references therein

[79] A.C. Legon: Chem. Soc. Rev, **19**, 197 (1990), **21**, 153 (1991)

[80] S.A. Cooke, G.K. Corlett, D.G. Lister, A.C. Legon: J. Chem. Soc. Far. Trans. **94**, 837 (1998)

[81] F. Lahmani, A. Zehnacker-Rentien, M. Broquier: J. Photochem. Photobiol. **154**, 41 (2002)

[82] R. Knochenmuss, I. Fischer: Int. J. Mass. Spectrosc. **220**, 343 (2002)

[83] O. David, C. Dedonder-Lardeux, C. Jouvet: Int. Rev. Phys. Chem. **21**, 499 (2002)

[84] C. Crépin, A. Tramer: Chem. Phys. **156**, 281 (1991)

[85] D.F. Kelley, E.R. Bernstein: J. Chem. Phys. **97**, 3841 (1992) and references therein

[86] H.Miyasaka, A.L Tabata, S. Ojima, N. Ikeda, N. Mataga: J. Phys. Chem. **95**, 8222 (1991)

[87] H. Miyasaka, A.L. Tabata, K. Kamada, N. Mataga: J. Am. Chem. Soc. **115**, 7335 (1993)

[88] J.A. Syage, J. Steadman: J. Chem. Phys. **95**, 2497 (1991) and references therein

[89] M. Rini, B. Magnes, E. Pines, E.T.J. Nibbering: Science **301**, 349 (2003)

[90] J.T. Hynes, T-H. Tran-Thi, G. Granucci: J. Photochem. Photobiol. **A 154**, 3 (2002)

Index

CPSIA information can be obtained at www.ICGtesting.com
Printed in the USA
LVOW01s1016040813

346172LV00008B/290/P

9 783642 064098